A Handbook of Northern France

by

William Morris Davis, S.D., Ph.D.

Professor of Geology, *Emeritus*, Harvard University
Professeur agréé à l'Université de Paris, 1911–12
Chairman, Geography Committee, National Research Council

Cambridge
Harvard University Press
1918

PREFACE

BY COL. PAUL AZAN

IF I were asked to draw up a list of the things that an American soldier, embarking for France, ought to carry with him, I should put at the head of it this little book, not only because the reading of it will be a pleasant distraction on the ocean voyage and in the trenches, but because a knowledge of its contents is indispensable.

The American soldier will certainly wish to know something of the region in which he is called to serve, to suffer, and perhaps to die. His friends also will desire to inform themselves about it. They will find in this Handbook a concise account of northeastern France, the equivalent of which can be learned elsewhere only by consulting a number of general works and special articles.

No one can be better qualified than Professor Davis to write such a book. His lectures on France, as a part of his course on Europe given at Harvard University from 1885 to 1912, have been based not only on a study of the best European sources, but on many journeys abroad, during which France has been repeatedly visited. Leading French geographers have honored him with corresponding membership in the Geographical Society of Paris, and in the Academy of Sciences.

If they had had to designate an American author for this Handbook, their choice would certainly have fallen on Professor Davis. His work in preparing the book is a service for which the United States and France should be equally grateful to him.

PAUL AZAN,
Lt.-Colonel, chef de la mission militaire française

CAMBRIDGE, August, 1917.

PREFACE

BY THE AUTHOR

THIS Handbook has been written with the approval of the Geography Committee, National Research Council, for those officers of our National Army who may wish to learn something of the leading physical features of the brave country where their aid will be so welcome. If the chapters are read over and the geographical names are identified on the accompanying maps, the uplands and valleys, rivers and cities, which may otherwise float vaguely "somewhere in France," will take proper position with respect to each other. News from the front and beyond and from the country that supports the front will thus become more definite and intelligible.

Much fuller information on French geography can be obtained from Commandant Barré's " La Géographie Militaire " (Paris, 1899), and more especially from his larger work, " L'Architecture du Sol de la France " (Paris, 1903). The historical aspects of the subject are admirably treated in the " Tableau de la Géographie de la France " (Paris, 1911), written by Professor Vidal de la Blache, the leading geographer of his country, as the first volume of the " Histoire de France " by Lavisse. For limited districts, reference should be made to

Auerbach's "Plateau lorrain" (1893), Demangeon's "Picardie" (1905), and Blanchard's "Flandre" (1906); but modern works such as these are unfortunately not available for all parts of France.

The geographical features of northeastern France and the adjacent regions are by no means so simple as those of an equal area of our prairie states. The dominant features — the "upland belts" — of the part of France here described are of a kind that is not common in the United States and hence not familiar to most Americans. They are of vital importance in warfare, as is shown in Johnson's recent and valuable book, "Topography and Strategy in the War." If our officers wish to know these features as well as they are known by the officers of the German army, they should study not only the condensed descriptions of such a Handbook as this, but all other available sources of information, particularly the large-scale maps that are accessible in France.

When the uplands and valleys of the country are known, it is a comparatively easy matter to locate the cities, villages, forests, railways, and roads of any district with respect to the relief of the surface; and when all these facts are learned, military movements may be planned with respect to them. It has not been possible, however, to indicate the roads and railways on the small outline maps which are here introduced as a means of locating the larger features of the region; and in order that the text shall not exceed a moderate number of pages, space has been allowed only for brief descriptions of a few of the most important lines of transportation.

Detailed information on these matters must be sought from special sources.

The descriptions here presented have been prepared with constant reference to the large-scale maps of the French "État-major" from which certain small rectangles are reproduced on a scale of 1:100,000. The generalized bird's-eye views sketched from these and other maps, will, it is hoped, assist the reader in visualizing the districts thus represented; the views are seen to best effect if the book is laid flat and looked at obliquely. It should be understood that these sketches omit a multitude of small features. Large-scale maps should always be consulted for details.

Certain sections of the introductory chapter as well as the whole of the final chapter have been revised by some of my colleagues, to whom I am much indebted. The interest in the book shown by the Officers of the French Army at Harvard has been a great encouragement; and for the prefatory page by Colonel Paul Azan I am under special obligations. To a number of friends who have contributed to the fund by which the publication of a first edition of the Handbook for free distribution to army officers has been made possible, my sincere thanks are given.

Many pleasant journeys have been recalled while writing these pages: early visits to France in 1868, 1873, and 1878; a bicycle tour across northern France in 1894; personal excursions in 1899, 1900, 1903, and 1905; and university excursions in 1908, 1911, and 1912. Now, at an age when travel is no longer so easy as it was once, the author can return only in imagination but always with deep sympathy to the fair landscapes, long familiar, so many of which have been laid waste. May the readers of the book come to share with the writer a warm affection for the scenes here described.

W. M. D.

Harvard University,
Cambridge, Mass.,
February, 1918.

CONTENTS

CHAPTER V

CHAPTER VI

CHAPTER VII

CHAPTER VIII

CHAPTER IX

A HANDBOOK OF NORTHERN FRANCE

CHAPTER I

A GENERAL ACCOUNT OF FRANCE

1. *France and its Central Highlands.* The parts of western Europe which through the course of centuries have been welded together to form the country we now know as France, the home of a brave people of an intense national spirit, may

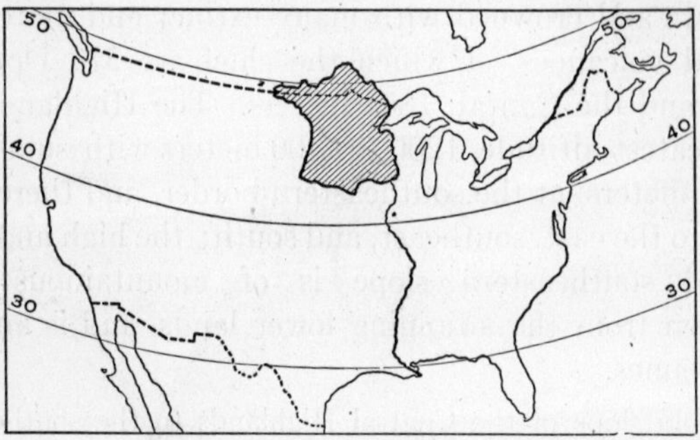

FIG. 1. FRANCE AND THE UNITED STATES

be described as an irregular rectangle, measuring 700 kilometers north and south by 600 kilometers east and west. If superposed on North America in proper latitude it would lie mostly in the northern United States. Its area (including the island of Corsica, 8747 sq. k.) is 536,400 sq. k. or 207,170 sq. miles, somewhat more than that of Indiana, Illinois, Wisconsin, and Iowa combined, or about midway between the areas of California and Texas. Its population has increased slowly from 37,386,313 in 1861 to 39,601,509 in 1911, and thus equals eight-ninths of that of the United Kingdom of Great Britain and Ireland, two-thirds of that of the German Empire, four-fifths of that of Austria-Hungary, and two-fifths of that of the United States.

The greater part of France consists of lowlands and moderately elevated uplands. Lofty mountains are found only in the Alps which form the southeastern, and in the Pyrenees which form the southwestern frontier. Between the northern lowlands and uplands and the lowlands of the south a gradual southeastward ascent leads to the Central Highlands or *Massif central*, flooded in certain districts with ancient lava flows and crowned with many extinct and more or less dissected volcanoes, of which the chief are Mt. Dore, 1886 meters, and the Cantal, 1858 meters. The Highlands reach their greatest altitude, 1200 to 1400 meters with summits up to 1700 meters, at the southeastern border, and there fall off rapidly to the east, southeast, and south; the high and deeply dissected southeastern slope is of mountainous aspect when seen from the adjoining lower lands, and is known as the Cévennes.

The bold slope of the Central Highlands to the southeast has exercised a marked influence on the history of France. When the Romans extended their Empire westward along the coast of the Mediterranean, they founded a province in the open country that is traversed by the south-flowing Rhone between the Alps on the east and the Cévennes on the west: this district is still called *Provence* and its language is not French but Provençal. The stream of invasion farther into Gaul was divided into two currents by the Cévennes: a smaller current flowed westward between the Central Highlands and the Pyrenees to the low plains of the southwest, now known as Gascony; a larger current flowed northward through the open valley of the Rhone to its continuation in the plain of the Saône, and thence northwestward over a saddle of higher ground to the extensive area of northern uplands and lowlands, now known as the Paris basin.

Some of the conquering invaders became colonists. The native Gauls gradually gave up their own Celtic language and adopted the Latin of their more civilized conquerors; but as their adopted speech had certain local peculiarities, the Romans called it *lingua gallica.* Although the smaller southern and larger northern areas of the region were confluent across a western lowland, they were elsewhere separated by the Central Highlands: hence, following a universal rule, the people of each area came to have certain ways of speech of their own. For example, in the south the habit was developed of using the Latin word, "hoc," pronounced *oc,* for "yes"; hence the southern language or Provençal came to be called *Langue d'oc,* and the southern district, *Languedoc.* In northern France, on the other hand, the affirmative was formed from the Latin words, "hoc ille," which in time came to be pronounced *oïl,* and was later reduced to the modern form *oui;* thus the speech of the northern region might be called *Langue d'oui.* It is chiefly the Central Highlands that are responsible for this linguistic division.

About the fifth century the Franks, a Teutonic tribe, overran the northern part of the country, subdued the inhabitants and adopted their language, which being thus further modified from the original Latin was called after the invaders, *lingua francisca.* As the northern region was much the larger of the two, it gathered the greater population, and the people of the south were in time dominated by their relatives on the north. There on a middle meridian, a quarter way from the northern to the southern limit of the country, Paris grew to be a great city, and the language of the north came to be the standard for the nation. Thus France today, peopled chiefly by the descendants of the original Gauls, of the Romans from the south, and of the Franks from the north — with the addition of a Norman stock in Normandy, and of Britons in Brittany — has taken its forms of speech from the southern invaders, but the name of the country and the name of its people and of its standard language come from the northern invaders.

It is curious to note that while the people of the country that thus gained the name of France call themselves and their language by the Latinized adjective, *français,* we follow the Franks in the Teutonic habit of changing the vowel in the substantive when making it adjective, and therefore the English name for the people and th language of France is French.

2. *Boundaries.* The boundaries of France may be de scribed in terms of their local departures from a rectangula

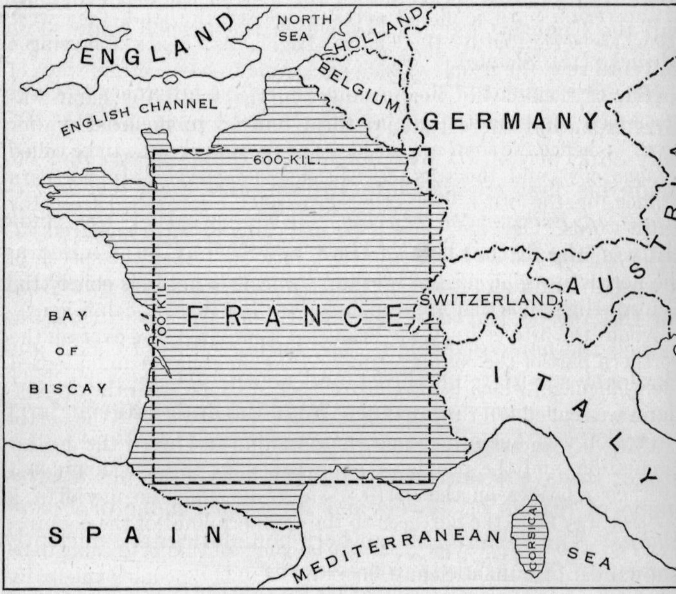

FIG. 2. FRANCE AND THE ADJOINING COUNTRIES

frame measuring 600 by 700 k. as here shown. The north ern boundary is bent outward 170 k. at the middle, so as t form a northern salient, close to the angle of which lies th city of Dunkirk on the coast of the North Sea. An irregula line trending southeast from the angle forms the boundary with Belgium across the low plains of the district known a

Flanders and along the southern slope of the gradually ascending highland of the Ardennes to the German frontier, described below. The southwestern line, following the low shore of Flanders for a short distance, soon reaches the clift coast of an area of uplands, and this is followed along the somewhat sinuous shore line of the arm of the sea which we call the English Channel, but which the French know as *la Manche* (the Sleeve).

The first salient of the sinuous shore line advances to Cape Gris Nez, and there reduces the Channel to its least width, only 33 k. across: it was hereabouts that Caesar, B.C. 55, made his first crossing into Britain, *quod inde erat brevissimus in Britanniam trajectus*. In fine weather the chalk cliffs on the farther side of the Channel may be descried as a whitish band along the horizon; and it is held by some that for this reason England, as viewed from the Continent, has gained the name of Albion. The harbor cities of Calais and Boulogne-sur-Mer, northeast and south of Cape Gris Nez, derive their chief importance from the international ferry traffic, day and night, across the Channel to Dover and Folkestone; thus the narrowed Channel here gains its English name of Straits of Dover, and its French name of *Pas de Calais*. The western and southern boundaries are sufficiently shown on the outline map on page 6.

On the east, the crest of the southwestern Alps, trending irregularly north–south, west of which the mountains extend over 100 k. toward the Rhone, forms the boundary with Italy, somewhat exterior to the southern third of the eastern side of the rectangle; the greatest excess is in the south near the coast. Mt. Blanc (4810 m.), the highest summit of the Alps, lies on the northern part of this line, beyond which French and Swiss territory interlock in such manner that the former occupies most of the southern side of Lake Geneva, while the city of Geneva is included near the end of a southwestern lobe of Switzerland that, obliquely limited by the northeast–southwest Jura highlands, sharply indents the eastern side of France.

3. *The Northeastern Frontier.* The northeastern boundary remains to be described. Beyond the other land frontiers of France, the near-by people of the adjoining nations, Spain, Italy, Switzerland, and Belgium, have many racial ties with the French; they speak languages of the same Romance family, and most of them (except in Switzerland), like most of the French, profess the Catholic religion. But beyond the northeastern frontier lies Germany, occupied by people of another stock, who speak an altogether different language and of whom the dominant members are Protestants. This frontier has for centuries been a battle ground.

Between the Jura and the Vosges, the boundary runs northward across a narrow depression, the passage of Belfort, whence the southwestward drainage runs by the Doubs and the Saône through the mid-eastern lowland of France to the south-flowing Rhone; while the northeastward drainage runs by the Ill to a similar lowland in western Germany, northward through which flows the Rhine. It is from here north- and northwestward that the frontier was set back from its former position after the Franco-Prussian war of 1870. Beyond the low passage of Belfort the boundary now follows the north–south crest of a short mountain range, known as the Vosges (German, *Vogesen*), for nearly 100 k., and thus the province of Alsace, extending into the lowland of the middle Rhine east of the Vosges and including the famous city of Strasbourg (German, *Strassburg*), was transferred to Germany under the name of Elsass. Northwest from the Vosges the boundary runs arbitrarily across the uplands to the little Grand Duchy of Luxembourg on the southern slope f the Ardennes highland, traversing the hills and valleys, first n the east then on the west of the north-flowing Moselle, with ttle regard to local features; thus the French province of ,orraine was divided, and a part of it, including the city of Ietz on the Moselle north of the confluence of the Meurthe, vas transferred to Germany under the name of Lothringen. 'he Luxembourg frontier, narrowed to less than 10 k. where t faces France, is followed by that of Belgium along the southrn slope of the Ardennes, beyond which we return northvestward across the lowlands to the coast near Dunkirk, vhere this description began.

Back of the disputed northeastern frontier the French stablished the strong fortresses of Belfort near the boundary n the passage south of the Vosges, Epinal on the western lope of the Vosges, Toul on the upper Moselle, Verdun on he middle Meuse, Mézières on the Meuse near its entrance nto the gorge by which it trenches the Ardennes, and Iaubeuge, near the border farther northwest; these forresses (the last two, unfinished) will be referred to again in he accounts of their several districts.

4. *Rivers and Cities.* The northeastern frontier of France s exceptional in being drained by rivers which flow through oreign territory on their way to the sea, as will be detailed arther on. This is nowhere else the case. A large northern rea — the greater part of the so-called Paris basin — is lrained northwestward to the English Channel by the Seine, vith Havre (population in 1911, 136,159) on the northern side of its estuary, Rouen (124,987) on its right or northern bank 70 k. inland, and Paris (2,888,110) on both sides of the river near the center of the basin, 170 k. from the sea; a number of branches — Oise, Aisne, Marne, Seine, Armençon, Yonne, and many smaller streams — converge

toward this center from the northeast, east, and southeast. In the plains east of Paris, drained by these rivers, the largest cities are Rheims (French, *Reims;* 115,178) and Troyes (55,486). The Somme, on which Amiens (93,207) is the

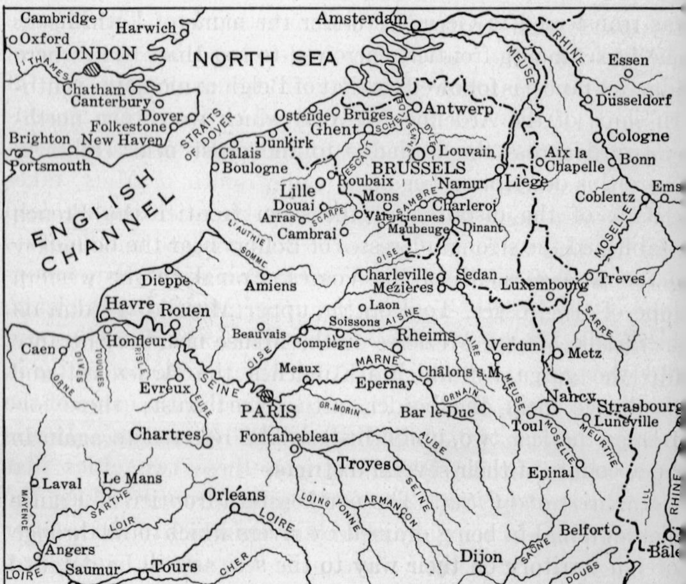

FIG. 3. THE RIVERS AND CITIES OF NORTHERN FRANCE

chief city, is the largest of several smaller rivers flowing northwestward to the Channel north of the Seine. The eastern part of the Central Highlands is drained northward through open valleys by the Loire and the Allier; these unite in a trunk stream which, under the name of Loire, turns westward and has the historic cities of Orléans (72,096), Blois (population of cities of less than 50,000 inhabitants is not given), Tours (73,398), and Angers (83,786) on its banks, and the port of Nantes (170,535) at the head of its estuary, which opens to the sea south of the peninsula of Brittany.

5. *Rivers of the Northeastern Frontier.* We now return to the northeastern border of France, where the French rivers traverse foreign territory in their lower courses. The Moselle and the Meurthe, flowing northwestward from the Vosges, join under the former name, with the important city of Nancy (119,949) on the Meurthe 10 k. above the confluence; then the single river turning northward and crossing the boundary in an open valley south of Metz, takes the German name of Mosel, turns northeast through a deep and winding valley, and joins the Rhine in the middle of its gorge through the broad highlands known as the Slate mountains (German, *Schiefergebirge*): at the junction lies Coblentz, a modernized form of the Latin name, Confluentia; and further down the Rhine lies Köln, which we know better in its French form, Cologne, the modernized Latin name of the Roman Colonia, established nearly 2000 years ago. It was from points on the Rhine between these two cities that German strategic railroads were built westward along the northern base of the Slate mountain highlands to the frontier of Belgium near Liège (German *Lüttich*) in the years preceding the war, evidently in readiness for use in the premeditated invasion of that neutral country.

The Meuse, rising in the hills west of the Vosges, flows north and northwest as an almost branchless trunk through a beautiful winding valley in the uplands between the Moselle on one side and the northeastern tributaries of the Seine system on the other; thus approaching the frontier on the hilly southern border of the Ardennes, the Meuse receives the Chiers from the east, below the junction of which lies

Sedan, and the Sermonne from the west, with Mézières Charleville on its bent course near-by: the river then turn north again and trenches the broad highland of the Ardenne in a deep and winding gorge, shifting the boundary 35 k northward with it; next bending northeastward along th northern slope of the Ardennes, and northward near the eastern border of Belgium, with Namur and Liège at the elbows it continues to and through Holland, where it is called th Maas, and finally joins the complex estuary of the lowe Rhine.

In the lower region farther northwest, the boundary is agai bent outward, but not so far as on the Meuse, at the crossin of three small rivers: the Sambre, the Escaut, and the Lys Here a number of details may be mentioned, because of thei importance in the war. The Sambre, flowing northeastwar from the rolling uplands of a salient French area in which lie the fortified city of Maubeuge near the point where Caesa "overcame the Nervii," crosses the border and runs along th northern slope of the Ardennes; Charleroi in the Belgian coa field, lies on its mid-length; it flows into the Meuse at th Namur elbow. The upper Escaut, on which Cambrai an Valenciennes are situated, is joined in the lowland near th boundary by the Scarpe, on which Arras and Douai ar placed; the Belgian city of Tournai is on the northwar course of the Escaut not far beyond the boundary; Mons lie in Belgium about midway between Tournai and Charleroi The Lys, flowing in another lowland northeastward across th border, receives some small branches from the south, on on of which lies Lille (217,807), with Roubaix (122,723) and Tourcoing (82,644) near-by in a famous industrial district Courtrai lies on the Lys a short distance in Belgium; farthe on, Ghent (*Gand*) is situated at the junction of the Lys with the Escaut, which then turning eastward with the name of Schelde, turns north again and then northwest; here it expands, with Antwerp (*Anvers*) at the head of the tide, into an estuary, the southwesternmost of several broadened coastal waterways, and reaches the sea by passing 50 k. through Dutch territory. Still farther toward the northern corner of France, the Yser, a small stream, flows eastward across the boundary, then northward to the coast; Ypres lies on one of its little branches in Belgium, and Nieuport marks its mouth in the dunes.

It is noteworthy that France has only five cities — Paris, Marseilles, Lyons, Bordeaux, and Lille — with populations over 200,000; and only ten more with populations over 100,000, of which Nantes, Havre, Rouen, Roubaix, Nancy, and Rheims are in the northern half of the country. The population of Paris is about as great as that of the fourteen other French cities which exceed 100,000.

6. *The Climate of France.* The climate of France is much more temperate than the climate of an area of the same latitude in central or eastern North America. The prevailing winds come from the west and bring with them the tempering influences of the ocean; moreover, they come somewhat from the southwest in winter and thus diminish the cold, and somewhat from the northwest in summer and thus moderate the heat which would otherwise be felt. The mean temperature in January (from 6° C. = 43° F. in the south to 2° C. = 36° F. in the northeast) corresponds to that of North Carolina and northern Georgia or of Arkansas and Oklahoma in the same month. Winter weather is frequently cloudy and wet; hence the air is chilling though the temperature is not very low. The coldest winter winds are from the continental interior on the northeast. The mean temperature in July (from 24° C. = 75° F. in the southeast to 18° C. = 64° F. in the northwest)

corresponds to the July mean of southern Pennsylvania and Ohio or of Wisconsin and North Dakota. The extremes of both seasons are less in France than in the central United States.

The annual rainfall varies from 500 to 1000 millimeters (20 to 40 inches), corresponding in general terms to that of eastern Nebraska and Iowa. Snowfall is rarely heavy, even in the north; and as the winds that follow snow storms usually come from the ocean at a temperature above freezing, snow seldom lies long on the ground. Weather changes, including the large cloudy areas of low barometric pressure with shifting winds and rain or snow, as well as the smaller thunder storms of summer, advance in a general way from southwest to north-east, as in the eastern United States; but the tracks of low-pressure centers, which often traverse the United States, usually pass to the north of France in spite of its relatively high latitude; hence France more often receives the southerly than the northerly winds that spiral around such centers.

The climate of northern and central France is fitting for wheat and other grains. Through the southeastern half of the country the vine is extensively cultivated and wine of many kinds is produced in great amount. In the extreme southeast the fig, the olive, and the orange flourish. Most of the common trees are of familiar kinds: they include oaks, maples, elm, beech, birch, chestnut, pine, ash, poplar, and willow. The *genet* or broom-plant is abundant in uncultivated fields; heather prevails in moorland districts.

7. *Government*. France was a kingdom for centuries under the Orleans and Bourbon dynasties before the First Republic was established, following the terrible revolution which began in 1789. The republic was replaced by the First Empire under Napoleon in 1804. In 1814 the kingdom was restored under the Bourbon dynasty, and continued with a short interruption

due to the return of Napoleon in 1815 (Louis XVIII, 1814, Charles X, 1824) until the revolution of 1830, when the Bourbons were replaced by Louis Philippe of the Orleans dynasty. Another revolution in 1848 caused the abdication of Louis Philippe and the institution of a second Republic, with Louis Napoleon Bonaparte, nephew of the first emperor, chosen as president by popular vote. Four years later the president of the republic took the title of Napoleon III, and the Second Empire was established. This lasted until the surrender of the Emperor to the Germans at Sedan in 1870, when a revolutionary government established itself in Paris and subsequently inaugurated the third Republic, which has now endured longer than any form of government since the old kingdom.

The Government thus constituted includes a Chamber of Deputies with nearly 600 members elected by universal (male) suffrage every four years, and a Senate of 300 members who are chosen by electoral colleges, consisting of local deputies and other officials, for nine years in three groups, one group every three years. Legislation may be initiated or amended in either body, but must be passed by both. The president of the Republic is elected for seven years by the senators and deputies in joint session, known as the constitutional assembly; these elections are not preceded by a popular campaign, but are accomplished promptly when a presidential term lapses, or is closed by resignation or death. The presidents of the third republic have been Thiers, 1871; MacMahon, 1873; Grévy, 1879; Carnot, 1887; Casimir-Périer, 1894; Faure, 1895; Loubet, 1899; Faillères, 1906; Poincaré, 1913.

The president is supported by a cabinet or ministry (*ministère*) the members of which (*ministres*) are selected by a political leader. The chief members are the ministers o finance, war, marine, interior, justice, foreign affairs, colonies public instruction, etc. The ministry represents the domi nant group or *bloc* of parties in the chamber of deputies; th ministry usually resigns when its policy is defeated by th chamber, and another leader is then selected by the presiden to form a new ministry. Thus the ministry does not repre sent the policy of the president, as the cabinet does in th United States, nor of a single political party, but of th people as reflected by a majority of the deputies, temporaril united in a *bloc* or coalition of several parties.

France was formerly divided into some thirty provinces, such as Provence, Gascony, Champagne, Normandy; it is now divided into eighty-seven departments, from which deputies are elected The departments are usually named after the rivers that they partly include, as Aisne, Oise, Seine et Marne, Meurthe et Moselle. The administration of each department is in charge of a prefect, who is appointed by the president of the republic on the recommendation of the minister of the interior. The departments are divided into 362 arrondissements, and these into 2915 cantons and over 36,200 communes.

The chief external possessions of France are Algiers and Tunis in north Africa; several large equatorial provinces in Africa south of the western Sahara and north of the Congo; Madagascar, the large island east of southern Africa; Indo-China, south of China proper; and New Caledonia and the Society Islands in the Pacific Ocean.

8. *Public Works: Roads.* The excellent highways of France (*Routes nationales, Routes départementales*) are among the most conspicuous of the many excellent products of the Department of Public Works. The pressing need of a great highway system in the United States can be better understood after a visit to France. The *Routes Nationales* are so planned as to provide thoroughfares connecting all the important centers

f population. They are carefully located and excessive radients are avoided; for over a century they have conributed greatly to the thrift of rural France. It not infreuently happens that a modern road follows, for a greater or ess distance, an ancient Roman road, traces of which may till be recognized even where it is abandoned. The less mportant local roads are in the care of the communes that hey serve.

Each main road has a name, as *Route de Paris à Nancy,* which is epeatedly given on signs along its course. Distances are accurately ndicated on the national roads by larger stones at every kilometer, nd by smaller stones at every tenth of a kilometer. Trim heaps of roken stone, ready for mending the surface, are — at least in times f peace — characteristic features of the unfenced roadside. Village ames are indicated, together with the department and arrondissenent to which they belong, on signs at the entrance of the main oads. The rule of the road is, as in the United States, " turn to the ight," and not as in Great Britain, " turn to the left."

Canals follow many of the larger valleys. They are so ocated as to afford communication not only along each main iver valley, but also over low passes between the valleys of eighboring river systems. They not infrequently pass under livides in tunnels two or three kilometers in length. Canals re supplemented by navigable rivers, improved by dams and ocks. Chains laid along the bed of certain rivers are passed ver a drum on tow-boats and are thus used to drag barges gainst the current. Although supplanted by railways as a neans of rapid transportation, the canals still have a large alue in keeping down the charges for slow freight.

Railways, some of which are owned by the State, are of two gauges: standard, 1.44 m. (4 ft. $8\frac{1}{2}$ in.) and narrow, 1 m.; the atter are for local service on branch lines of light traffic. Railways are so numerous that they cannot be shown on

the small maps in this book. The strength of bridge construc
tion over or under railways and the number and height o
railway viaducts are impressive. The frequent use of skewe
arches of stone or brick is surprising to visitors from a countr
where square bridges of wood, steel, or concrete prevail.

Passenger cars (*wagons*) for local service are of small length
divided into transverse compartments, entered at either side; i
trains of such cars one cannot pass from compartment to compart
ment, much less from car to car. On express trains, longer ca
connected by "vestibules" and divided into compartments with
corridor on one side, are generally used, but these also have doors o
both sides, sometimes for each compartment. The cars or compart
ments are of three classes; the class chosen (*première*, *seconde* (pron
segonde), *troisième*) must be specified when buying a ticket. Ticket
are usually examined as the passenger goes from the waiting roon
to the platform, and collected as he leaves the station of desti
nation.

Forests. Certain forests in France are under governmenta
care, as timber is of high value in a country of ancient occupa
tion. Many of the forested areas are located on uplands o
relatively infertile soil, or of so uneven a surface as to mak
their use difficult for other purposes than tree-growing
Most of the forests are traversed by many rectilinear lanes
laid out in geometrical pattern so as to intersect at selecte
centers, and used in exploitation of timber.

Harbors. Harbor works are extensive and elaborate. They
are to be seen at all the ports along the clift coast of Nor
mandy and Picardy, and along the dune-bordered coast o
Flanders, where they include jetties to prevent the closin
of the harbor entrance by the 'long-shore drift of sand an
gravel, and stone wharves adjoining dredged docks.

Topographic maps. The topographic map of the genera
staff, *Carte de France de l'Etat-major*, is the standard on whicl

all other modern maps of France are based. It is the work of army engineers begun in 1818, finished in 1866, revised in later years, and published by the Geographical Service of the Army in 274 sheets on a scale of 1:80,000 (8 kilometers = 1 decimeter, or about 1 mile = $\frac{3}{4}$ inch). Another edition of the same survey, revised, is published in quarter-sheets on a scale of 1:50,000. Selected rectangles from the 1:50,000 quarter-sheets are here reproduced on half-scale (1:100,000, or 1 cm. = 1 k.) on pages 60, 74, 78, 90, 94, 118, 126, 160; their location is shown by small rectangles, with page numbers, in the maps on pages 28, 54, 67, 134.

These maps are printed in black. Relief is well indicated by down-slope lines, or hachures, which are drawn short, heavy, and close-set for steep slopes, long, fine, and open-spaced for gentle slopes. Flood-plains subject to overflow are stippled; small streams are shown by single waving lines; larger streams by double lines with the included space shaded: forests, nearly all of which have special names, by conventional tree-signs; roads, by double lines with unshaded space between, three grades of importance being indicated by the width of the space (rows of trees adjoining the national roads are shown by dots); lanes, by single lines; paths, by dotted lines; railways by single heavy lines; canals, by three close-set lines, the middle line heaviest. Altitudes are given in meters for occasional points. Cities and villages are shown with the pattern of their chief streets. Boundaries of departments are marked with strong short-dash lines, the seat of their prefecture being marked with a small rectangle containing the letters PF; boundaries of cantons are marked with dotted lines, the name of their chief town being adjoined by a small oval containing the letters CT.

Maps based on a later survey along the northeastern frontier have been prepared on a scale of 1:20,000, with altitudes shown by contour lines of 5 m. interval; these sheets are printed in several colors; they are not sold to the public.

Reductions of the standard 80,000th map have been published on smaller scales and in various styles by several departments of the government. Those prepared by the Geographical Service of the Army and by the Ministry of Public Works are both on a scale of 1:200,000; the first in 78 sheets, the second in 135. The maps of the Geographical Service are very legibly printed in five colors; the relief in brown shading with 20 m. sketched contours, the water in blue, forests in green, names and railways in black, towns and highways in red. These maps are the most serviceable for general use.

9. *Money.* French money is reckoned in francs and centimes. A franc is normally worth $0.19; five centimes or a sou is about equivalent to a cent; 10 centimes or two sous, to an English penny. Five francs are almost the same as a dollar, and 25 francs are closely equivalent to a pound sterling.

French coins are: copper; 5 centimes or 1 sou; 10 centimes or 2 sous.
silver; 50 centimes = 10 sous or half a franc; 1 franc; 2 francs; 5 francs.
[NOTE: Many Belgian and Italian silver coins, nominally equivalent to French money, do not pass at their face value.)
gold; 10 francs; 20 francs = a napoleon.

English money: 12 pennies = 1 shilling = $0.24.
20 shillings = 1 pound = $4.84.
(21 shillings = 1 guinea).

English coins are: copper; half penny; penny.
silver; three pence; six pence; shilling; two shillings (florin); two and a half shillings (half crown); five shillings (crown).
gold; 10 shillings (half sovereign); 20 shillings (sovereign).

Weights and Measures. The decimal system of weights and measures, adopted by the French near the close of the eighteenth century and by most countries of Europe since then, is based on the unit of linear measure, or meter (*mètre*), which was intended to be and is almost exactly 1/10,000,000 of the meridian quadrant of Paris. It equals 39.37 inches or 3.28 feet. Its multiples and fractions and their equivalents are: —

Linear measure.	Meters.	Inches.	Feet.	Miles.
kilometer	1000	....	3281	0.62138
hectometer......	100	....	328.1	......
decameter.......	10	....	32.81	......
meter...........	1	39.37	3.281	......
decimeter.......	.1	3.94	0.328	......
centimeter	.01	0.39	0.033	......
millimeter.......	.001	0.04	0.003	......

One kilometer is roughly $\frac{3}{5}$ or $\frac{5}{8}$ of a mile.

Areas are expressed in square meters = 1550.0 square inches, or 10.76 square feet; in ares = 100 square meters; and in hectares = 10,000 square meters = 2.471 acres; hence 260 hectares about equal one square mile.

Volumes. The unit of volume is a cubic decimeter, called a liter (*litre*) = 61.02 cubic inches = 1.06 U. S. quarts = 0.88 British quart.

Weights. The ordinary unit of weight is the kilogram, which is the weight of a liter of water under standard conditions = 2.205 pounds.

Temperatures. The Centigrade thermometer scale has 0° at the freezing point and 100° at the boiling point of water, under standard conditions. To convert Centigrade into Fahrenheit degrees, multiply by $\frac{9}{5}$ and add 32°. To convert Fahrenheit into Centigrade degrees, subtract 32° and multiply by $\frac{5}{9}$.

Atmospheric Pressure. French weather maps represent atmospheric pressure in millimeters: normal pressure is taken to be 760 mm. = 29.92 inches.

CHAPTER II

THE GEOGRAPHICAL FEATURES OF NORTHERN FRANCE

10. *The Paris Basin.* The greater part of northern France is occupied by the so-called Paris basin, which gains its name

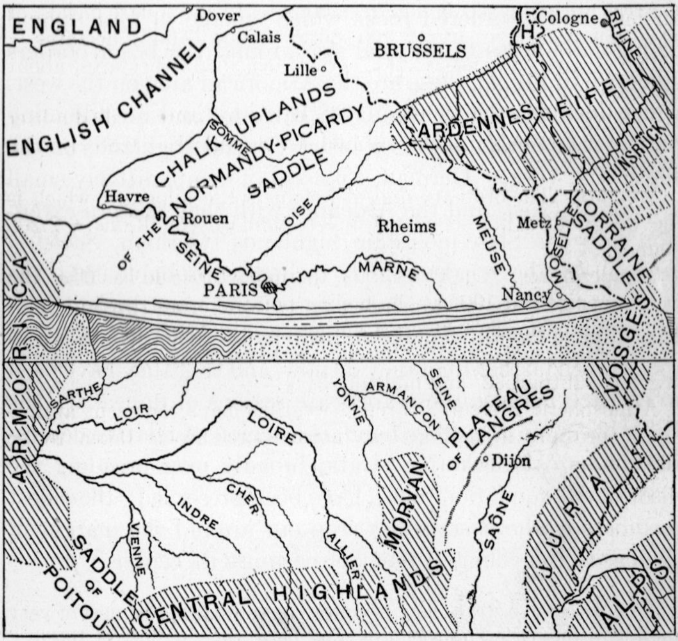

FIG. 4. THE PARIS BASIN AND ITS SADDLES

from the basin-like slope of the rock layers from all sides towards Paris as a center. The relation between the rock layers and the surface forms here occurring is moreover in many ways so manifest and so significant, that an understanding of it aids the memory in placing a multitude of details in their proper position with respect to the larger features of which they are parts.

The Bordering Uplands and Highlands. The stratified formations occupying the Paris basin lie, with a total thickness of hundreds or thousands of meters, upon a foundation of ancient and disordered rocks which emerge in four upland or highland areas of unequal size around the basin borders as shown in Fig. 4: these are the Armorican area on the west, which includes the peninsula of Brittany and an adjoining part of the mainland, the extensive Central Highlands on the south, the Vosges (German, *Vogesen*) of comparatively small area on the east, and the Ardennes with their eastern extension into the Slate-mountain highlands (German, *Schiefergebirge*) on the northeast. It is highly probable that the strata of the Paris basin, shown in section across the middle of Fig. 4, once overlapped the four areas of ancient foundation rocks much farther than they do now, and that they have been worn back because those areas are regions of upheaval; the Paris basin, on the other hand, is a region of relative depression, where the covering strata, broadly overspreading the disordered foundation rocks, have been preserved; the oldest members of the basin series crop out around the margin of the basin, the youngest member occupies its center.

The successive strata may therefore be compared to a nest of very shallow dishes, the largest one at the base, the smallest at the top, yet so nicely fitted together that the edges of all rise to about the same altitude. But the basin structure has many irregularities: Paris lies near its center, and the edges of the successive formations are farther apart on the east side than on the west; indeed, the basal members of the series that are broadly exposed along the border of the Vosges and in Lorraine are hardly seen elsewhere; and to the north, the lower members are overspread by an upper member — the chalk — so far that they are concealed as it laps upon the foundation rocks which ascend gradually eastward in the Ardennes.

11. *The Four Saddles.* Moreover, as may be seen in Fig. 4, the basin strata extend outward across the four depressions or "saddles" that separate the four enclosing uplands of ancient rocks. A broad and flat northwestern saddle, forming the chalk uplands of Picardy and Normandy, occupies the 330 k. space between the Ardennes highlands on the north of the basin and the hilly Armorican area on the west; the northwestern side of this saddle is cut off by the sea. A narrow southwestern saddle, about 60 k. across, lies between the Armorican area on the west and the Central Highlands on the south, and thus connects the Paris basin with the lowlands of Gascony: this may be called the Poitou saddle, after the old province of that district. A broader southeastern saddle, 150 k. in width, rises gradually and forms the so-called plateau of Langres, between the Morvan, a northeastern extension of the Central Highlands, on one side, and the much smaller mountainous mass of the Vosges on the other; its steeper southeastern side, the southern part of which is known as the "Golden Slope" (*la Côte d'Or*), with Dijon near its base, descends rapidly to the flat basin of the Saône, known as the plain of *la Bresse.* The fourth and northeastern, or Lorraine saddle, broadly exposing the lowest members of the basin series, has a width of 75 k. between the Vosges on the east and the Slate-mountain highlands on the north; this saddle stretches eastward into Germany and, rising gradually, is obliquely cut off by the broad valley-lowland of the middle Rhine; its eastern upland border, prolonged northward from

the Vosges and overlooking the Rhine lowland, is known as the Hardt.

It should be borne in mind that the Seine and its branches drain only a part of the Paris basin. In the northwest, several small rivers — the Somme and others on the northeast of the Seine, the Risle, Touque, Dives, and Orne on the southwest — have independent courses to the Channel. In the northeast, the Meuse and the Moselle flow out from the border of the basin through the adjoining uplands to the Rhine. In the south, the Loire, after emerging from the Central Highlands, turns westward across the southern part of the basin, receives certain tributaries — chiefly the Loir and the Sarthe — from the western part of the basin, as well as others from the south, and reaches the sea south of Brittany.

12. *Products of the Highlands and the Basin.* The Paris basin, differing from the enclosing uplands and highlands in the composition and attitude of its rocks, therefore differs also in form, soil, and mineral products. The uplands and highlands consist of granite, gneiss, schists, and other crystalline rocks, generally resistant to weathering, as well as of various stratified rocks, greatly deformed, much more ancient than those of the Paris basin, and usually more indurated. Coal is found within the boundaries of France only in these ancient foundation rocks: the Central plateau includes the important though small coal basin of St. Etienne in a valley that indents the middle of its eastern side, southwest of Lyons; not far north in another highland valley is Le Creuzot, with its great iron works. Another important coal area, known as the Sarre (German, *Saar*) basin after the branch of the Moselle which it borders, lies in German territory, as shown on p. 158, south of the Slate-mountain uplands by which the Lorraine saddle is limited on the north. The extensive coal basin of Belgium, shown on p. 153, lies along the northern side of the Ardennes, and extends westward under the cover of the overlapping basin strata into northern France, where its deep mines determine the situation of a number of industrial cities, of which the chief are Douai and Lens. Iron ore occurs in several districts of the ancient foundation rocks, especially in the northern arm of the Armorican area, known as the Cotentin; hence important iron works have been established near by in the ancient city of Caen.

The strata of the Paris basin include limestones, chalk, marls, sandstones, and clays, all lying nearly horizontal. Certain members of the series contain important iron ores; the chief of these lie in the uplands west of Metz, and largely in a part of Lorraine that was taken by Germany in 1871 (see p. 77). But as a rule the stratified formations yield few important mineral products, apart from building stones, limestone, cement, gypsum, and road metal. The best building stones are limestones and sandstones, which are easily carved when fresh from the quarry, but which become hard and durable after exposure to the weather. Flint concretions from the chalk provide a resistant road cover.

As the basin strata outcrop in northeastern France in roughly concentric arcs, or aureoles, Fig. 13, around Paris as a center, the varied forms of the surface, modelled by the action of erosional processes on the nested strata of varying resistance, as well as the soils and with them the agricultural products of the basin area, are closely sympathetic with the patterns of the concentric structural arcs; likewise, local industries as affected by soils and products, drainage lines and transportation routes as affected by surface forms, and population, both rural and urban, as affected by all these elements, are repeatedly found to be influenced if not controlled in their distribution by the same structural factors, as will be fully shown on later pages.

CHAPTER III

THE REGION AROUND PARIS

13. *The Three Sectors centering at Paris.* The confluence at Paris of the Seine and the Marne, each of which has received many converging tributaries in its upper course, may be taken to mark the drainage center of the Paris basin. The same two branch rivers and the trunk river in which they unite serve to divide the central area of the basin into three unequal sectors: one of about a right angle and a half on the south, between the upper and lower Seine; another of the same amplitude on the north between the Marne and the lower Seine; and a third of about 90° on the east between the Marne and the upper Seine. Although the sectors are here named after three of the cardinal points, it should be noted that the course of the upper Seine is about northwest; that of the Marne, west-southwest; and that of the lower Seine, west-northwest; hence the sectors do not precisely face the points for which they are named.

The southern sector is largely occupied by uplands, which are chiefly formed of the youngest, uppermost members of the heavy series of basin strata, lying essentially horizontal. Some 60 k. or more to the south and southwest these strata have a broadly continuous surface; near Paris they are separated by irregularly branching valleys into discontinuous tabular masses at remarkably uniform altitudes of 160 or 170 m. The eastern sector begins as a lowland 80 or 90 m. in altitude and ascends slowly eastward with the rising strata

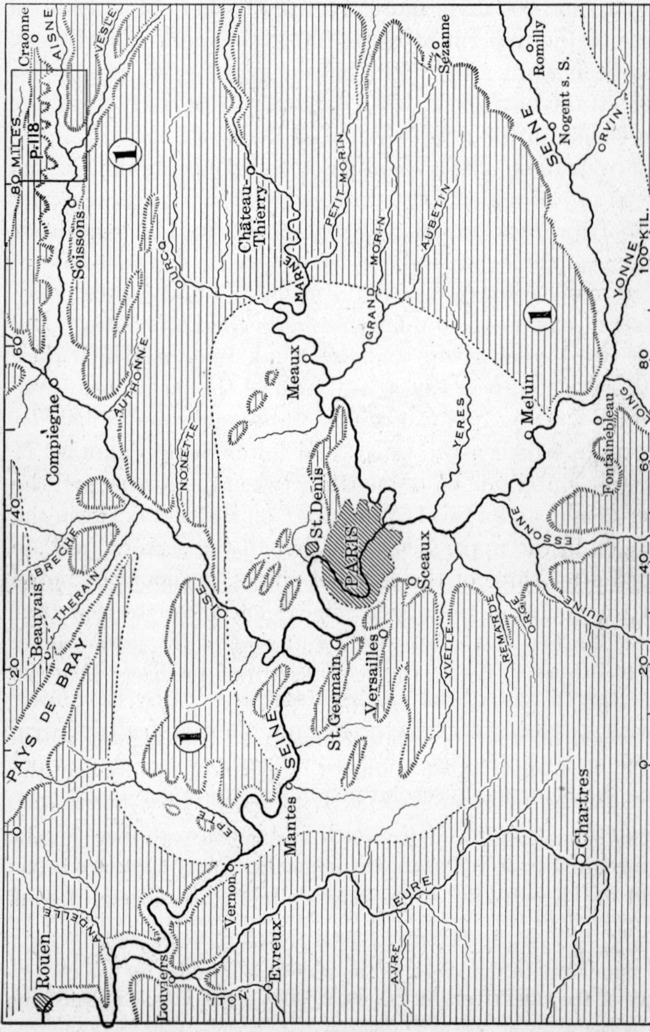

FIG. 5. CENTRAL AREA OF THE PARIS BASIN

to an upland 200 m. or more in altitude, where practically all traces of the uppermost beds, 100 m. or more in thickness, have been worn away. In the northern sector, the uppermost strata have been less completely removed; they are seen chiefly in the neighborhood of Paris as isolated residual hills, from 130 to 200 m. in altitude, surmounting the lowland (60–90 m.) of underlying strata, which rise slowly westward and northward to an upland (140–170 m.) that is continuous,

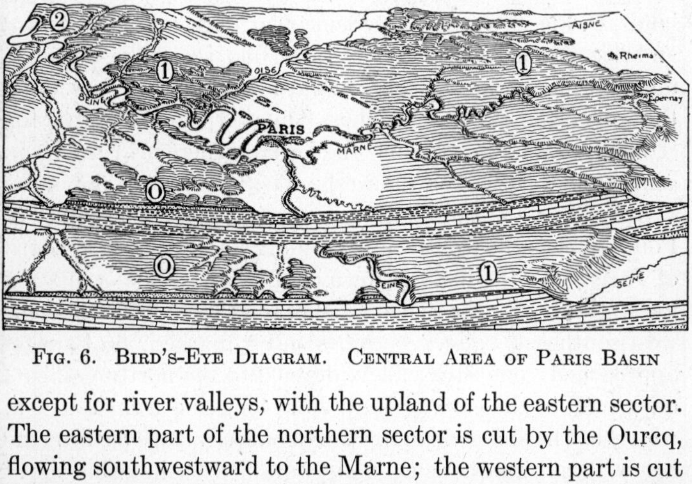

FIG. 6. BIRD'S-EYE DIAGRAM. CENTRAL AREA OF PARIS BASIN

except for river valleys, with the upland of the eastern sector. The eastern part of the northern sector is cut by the Ourcq, flowing southwestward to the Marne; the western part is cut across on a parallel line of greater length by the Oise, which joins the Seine at its second northern loop below Paris.

This arrangement of uplands and lowlands is very roughly summarized in Fig. 6, by which the unsymmetrical "nesting" of the basin strata, already referred to on an earlier page, may be made more apparent than before. It thus appears that the center of the Paris basin, as marked by the uppermost members of the basin series, lies somewhat to the south of the center as marked by the convergence of the chief rivers at the site of the capital city. The division of the Parisian district into sectors, as here indicated, is not exact and geometrical, but rough and geographical, for the three chief rivers are not rectilinear. The upper Seine is moderately sinuous; the Marne is strikingly sinuous; the lower Seine is exceedingly sinuous, especially at and next below Paris. Near the central area the rivers are about 20 m. above sea level; the adjoining lowlands are from 20 to 60 m. higher. The three sectors may now be described more in detail.

14. *The Southern Sector.* The discontinuous tabular uplands, O, Fig. 6, south of the Seine near Paris fall off eastward and less distinctly westward in irregular scarps, notched by many valleys and ravines, toward lower lands bordering the upper and the lower Seine. The scarp to the east is the stronger of the two, because the lowland there near the basin center is lower than on the west. The valleys that notch the scarps become shallower toward their heads. The forest of Fontainebleau lies on the upland margin (130 m.) 60 k. south of Paris; and the historic town of the same name occupies a lowland site near the Seine. A slight predominance of northwest-southeast valleys becomes more marked as Paris is approached: one such valley, drained to the northwest, cuts off a strip of the even upland, 25 k. long and from 2 to 4 k. wide, that lies tangent to the three southern loops of the Seine below Paris; another such valley, drained to the Seine above Paris, almost cuts off a shorter, irregular portion of the upland near the blunt apex of the sector.

The location of cities and towns near Paris on the south may be roughly expressed in terms of the features thus described. Sceaux lies on the high ground near the east end of the shorter detached portion of the upland, and enjoys a delightful prospect over the lowland adjoining the upper Seine; Meudon lies on a northern spur of the same upland detachment, and overlooks the first southern loop of the Seine below Paris, across which the Bois de Boulogne is seen covering the terminal part spur of the lowland that enters this loop from the northeast. Sèvres lies on the slope near the Seine below Meudon. Versailles (60,458) with its royal palace and gardens is situated near the head of the valley which cuts off the longer upland strip; St. Cyr, the seat of a famous military school, is at the northern base of the large upland area a little farther west. St. Cloud and its park occupy the eastern end of the longer detached upland strip, with a fine view up the Seine to Paris; St. Germain is at the northern base of this strip farther west, where it is touched by the second

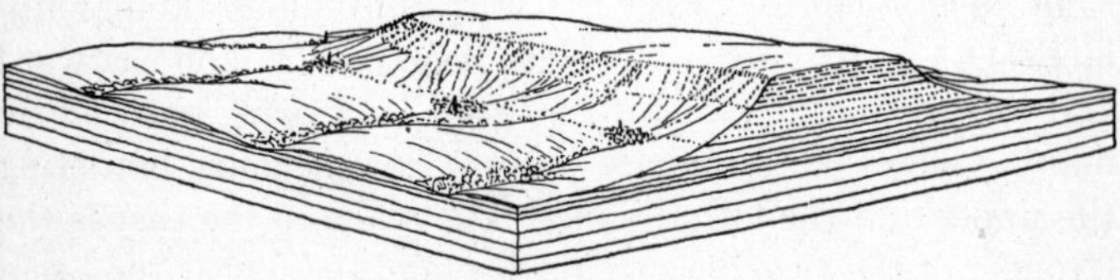

FIG. 7. VILLAGES AT STREAM-HEAD SPRINGS

southward loop of the Seine west of Paris. In this area of detached uplands, the smaller villages, most of which are of ancient or medieval origin, are frequently situated, as in Fig. 7, on the upland slopes where the higher beds of pervious limestones and sandstones rest, all horizontal, on impervious clays; for at that level springs issue, on which the local water supply has for centuries been dependent. This primitive control of village location will be frequently met with in other parts of France.

15. *The Eastern Sector.* The lowland of the 90° sector between the Seine and the Marne rises eastward to an upland, 1, Fig. 6, known as *la Brie*, part of which was overrun by the German army in August, 1914, as will be described on p. 114. The slope of the surface is, although gentle, distinctly greater than the fall of the rivers that drain it; hence their valleys increase in depth as they are followed upstream. Most of the streams evidently had a serpentine course before they began to erode the present valleys, for the valleys themselves are

strikingly sinuous, with steep amphitheatral slopes, alternately on the right and left, opposite sloping spurs, alternately on the left and right. As the serpentine curves of rivers increase in size with increase of river volume, the valleys show the same systematic variation of form; thus the Yères, a small

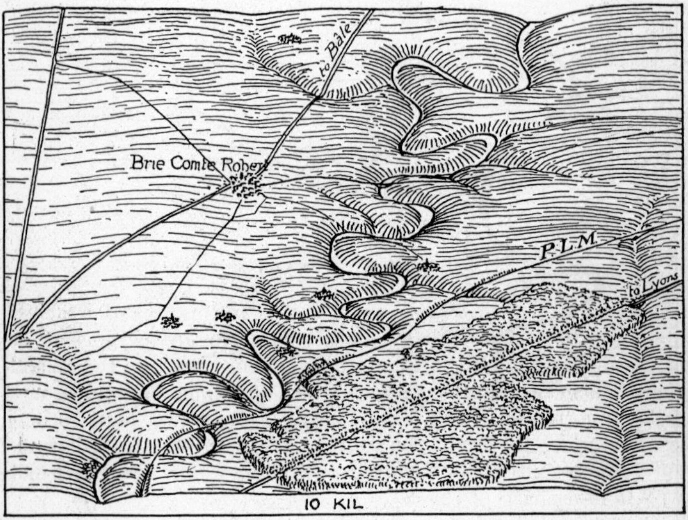

FIG. 8. THE MEANDERING VALLEY OF THE YÈRES, LOOKING EAST

tributary of the upper Seine, has a valley with numerous, small, close-set bends as in Fig. 8; the Grand Morin, a somewhat larger tributary of the Marne, has a valley of larger pattern and therefore of fewer bends; and the Marne itself has a valley in which the bends are on a still larger scale and hence still fewer in number, as in Fig. 9. The city of Meaux, here shown, marks the nearest approach of the Germans to Paris in September, 1914.

A national highway and a main railway — Chemin de Fer de l'Est — which ascend the valley of the Marne on the way from Paris

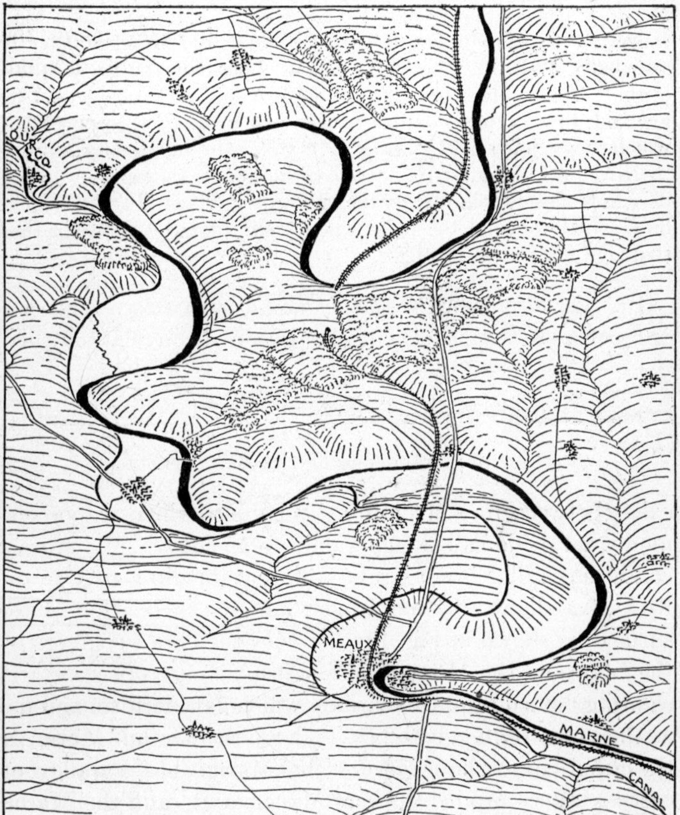

FIG. 9. VALLEY OF THE MARNE AT MEAUX, LOOKING EAST

to Nancy, make short cuts across the valley-side spurs; the highway is the shorter of the two lines, because it can follow steeper gradients than the railway; both are much shorter than the river, as is shown in Fig. 9. It was across this sinuous valley that the French under General Joffre drove back the Germans in the Battle of the Marne in September, 1914. Another highway and a main railway — Chemin de Fer de Paris à Lyon et à la Méditerranée, the so-called "P. L. M." — turn from the upper Seine valley into that of the Yères, in order to make a direct cut southeastward across the lowland, thus saving a detour to the west made by the Seine itself, which they rejoin about midway in its valley through the upland, and there Melun is situated; farther upstream, where the valley is deeper, it is more closely followed by the highway and the railway. The two lines a little beyond the point where they leave the Seine for their short cut are shown in Fig. 8.

16. *The Strata and the Soils of the Eastern Sector.* The eastern sector offers an excellent illustration of the relation which

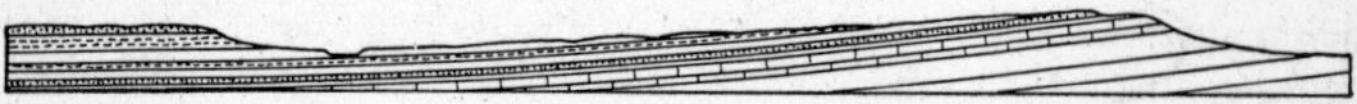

FIG. 10. STRUCTURE OF THE BRIE UPLAND

exists between rock structure — a subject which is too commonly set aside as belonging only to geology — and surface conditions. The basin strata in this sector dip gently to the west at a slightly steeper angle than the inclination of the surface, as shown in Fig. 10; hence as they are cut across by the Seine and the Marne, their occasional outcrops appear in slanting belts on the valley sides, which vary in form and soil as they pass from belt to belt. The most resistant strata are the impure limestones (*calcaire grossier*) which form the strong east-facing escarpment where the upland of Brie falls off toward the lowland of Champagne, as will be further described in chapter VIII.

It is noteworthy that, as a result of the strata being inclined at a steeper angle than the gentle descent of the surface as shown in Fig. 10, their uppermost (youngest) members in this sector are reached on the lowest ground, while the lowermost (oldest) members of this district occupy the upland — but as will be seen later, still lower members occur in the lowlands farther east. As the successive strata overlap the sloping surface, the soils vary, and with variation of soils comes variation of living conditions. The strata that overlie the *calcaire grossier* and cover most of the Brie uplands weather into a poor, "cold" or wet soil, which was formerly left to forest growth, and which even where cleared and cultivated today does not give so good a return as the calcareous soils of the lower slope or as the loams of the lowlands farther west.

These more fertile lowland areas have long been known for their thriving farms, each having its group of buildings enclosed by a rectangular wall, half a mile or more from its neighbors, instead of being compacted in villages such as are described above as having grown around the springs in the slopes of the uplands on the south of Paris, or such as will be described in section 56, grouped around the deep wells of the northwestern chalk country. It must be remembered that these varied relations between geographical factors and human conditions have long been well established in a country of so ancient settlement as France, where the manner of living and the location of settlements have been developed by primitive methods of trial adapted to simple, local needs through centuries of struggle for existence, and where satisfactory locations and occupations, when found, have been long adhered to. French geography is therefore in this respect utterly unlike the geography of the western United States, where the location of many a village has been arbitrarily determined by conditions little related to geographical factors.

17. *The Northern Sector*. The broad sector north of Paris combines some of the features of its two neighbors. Its central part is chiefly a rolling lowland, similar in form and soil to the lower part of the eastern sector; but the lowland is here sur-

mounted by a number of hills or upland remnants, mostly of small area, although in composition and altitude similar to the large uplands of the southern sector. One of the smallest of the hills, of less height than the larger ones, is Montmartre, in the northern part of Paris; a chain of hills, about 20 k. in length, begins in the Butte Chaumont on the northeastern border of Paris and extends eastward in a rambling course to the north bank of the Marne. A range of hills begins 35 k. northeast of Paris and trends northwest; the southeastward prolongation of this line into the valley of the Marne marks the site of Meaux. A small ridge (170 m.) is tangent to the two northward loops of the Seine, 15 k. northwest of Paris; north of this ridge rise the subdivided heights of Montmorency (182 m.), of oval area, 9 k. in length, 18 k. north-northwest of Paris; the little upland of Hautie (170 m.) of similar dimensions stands 30 k. to the northwest of the city, a short distance west of the confluence of the Seine and Oise, across the third southern loop of the trunk river below Paris. Nearly all the isolated hills near Paris are crowned by fortresses and redoubts, formerly regarded as impregnable but now known not to be of sufficient strength to withstand bombardment by the heavy siege guns of modern warfare.

The far northeastern uplands of the northern sector, from which the Germans were driven after the Battle of the Marne in September, 1914, and the northern extension of the sector in irregular, plateau-like segments beyond the east–west valley of the Aisne, where the battle front lies in the spring of 1918, will be described in chapter IX. The present account will not go beyond the upland area, known as Valois, which is limited on the north by the Aisne valley. The gradual ascent from the lowland (69–90 m.) near Paris to the Valois uplands (150–170 m.) is accompanied by changes in soil similar to those described for the eastern sector; but while the ascending surface was there comparatively continuous, except for a moderate number of radial valleys, it is here more interrupted by many rambling valleys, drained by branches of the Ourcq, Aisne, and Oise, as shown in Fig. 5.

The upland is furthermore occasionally surmounted by hills, the most important of which constitute a narrow, east–west range about 13 k. south of the Aisne valley: it is 35 k. in length and from 200 to 250 m. in altitude, and is cut through by a small stream at mid-length; the eastern half is mostly cleared; much of the western half is covered by the great forest of Villers-Cotterêts, which also spreads across the uplands to the north and south; like most of the larger forests of France, this one is traversed by a system of lanes arranged in geometrical pattern, chiefly for use in management and exploitation.

The northern margin of the upland overlooking the Aisne valley, like the margin of the rambling valleys which dissect the upland, is irregularly indented by the branching ravines of many small side streams. The general accordance of the upland level across the valleys, as well as the correspondence that may be noted in the rock layers on the two sides of the valleys, proves clearly enough that the upland would be a continuous surface but for the work of weather and streams in eroding valleys beneath it. This is a point of practical importance, since it shows that when one valley-bounded segment of the upland is known, the neighboring segments may be known in a general way from it, for they all resemble one another in essential features, such as altitude, structure, and general pattern, although they differ from each other in individual features, such as extent, and the number and direction of indenting ravines. It is the uncounted individual variations thus played on a simple scheme that characterize the Valois landscape.

18. *The Western Salients of the Northern Sector.* The northwestern part of the northern sector west of the Oise may be named after the district of Vexin, which it includes. Its western border resembles the eastern border of the eastern

sector in being determined by the gradual ascent of the relatively resistant *calcaire grossier*, which terminates in an escarpment overlooking an exterior lower land of weaker layers; but in the Vexin on the northwestern side of the Paris basin the ascent of the strata and the outlook of the escarpment are to the northwest, while in the Brie on the other side of the basin center they are to the east. The west-facing Vexin border is more irregular than the east-facing Brie border; its uplands advance westwards in several salients between lowland reëntrants, and the advancing salients are cut off in separate upland areas of different size by the valley of the Oise, which crosses the lowland reëntrants, as shown on the map, p. 28. As a result the Oise valley varies greatly in width and quality; it is rather narrow and well enclosed where it transects the resistant limestone strata of the upland salients; it is broadly open where it traverses the weaker strata of lowland reëntrants. Here is the most open approach to Paris.

The first of the reëntrants occurs just below the confluence of the Aisne and the Oise; it is occupied to the east of the Oise by the forest of Compiègne, called after a city of that name near the confluence. To the southwest of Compiègne, two contiguous cut-off salients (110–150 m.) divided by the river Brèche, advance northwestward; Clermont lies on the eastern border of the larger one. Another lowland reëntrant is crossed by the Oise half way from the Aisne to the Seine; this will be referred to later in the account of the Pays de Bray, p. 140. The next salient forms a large upland which spreads southward to the Seine; its westernmost spurs (160 m.) are cut off from the main area by the south-flowing Epte. The upland of this salient is unlike the others in being surmounted by several small hills (190, 200 m.), like those near Paris but of greater altitude because the platform from which they rise is higher than the Paris lowland.

19. *Paris and its Neighborhood.* The military visitor to Paris may find difficulty in selecting among its many attractions the few to which his limited time can be best devoted. If he passively follow

the conventional guidebook, he will be led, whether his tastes are artistic or not, to matchless galleries of painting and sculpture; or to famous buildings and monuments, whether or not he knows enough of history and architecture to appreciate them; but if his tastes are geographical, he will do well to select certain districts of medieval and modern Paris as samples for outdoor study. These should include parts of the old central city of narrow, irregular streets, and sections of the older and newer ramparts, marked by the ring of inner boulevards and of outer dismantled fortifications. A number of new avenues, cut through the older parts of the city, are striking and characteristic features.

Excursions to high points of view are next to be commended; first, in the city, to the Butte Chaumont on the east, to Montmartre on the north, and to the Arc de Triomphe and the Eiffel Tower on the west; the wide prospect from the top of the Arc, if permission be secured to make the ascent, is especially to be commended, with its view eastward down the finest avenue in the world — the Avenue of the Champs Elysées — to the Garden of the Tuileries, and westward to Neuilly and the Bois de Boulogne and across the loop of the Seine. Later, visits should be made outside of the city to the heights on the south near Meudon, and on the west above St. Cloud. A trip on the river is to be recommended; here a hand map is essential, in order not to lose one's bearings on the many river turns.

The different parts of Paris and the towns of familiar names near by, Fig. 11, are best learned in relation to the loops of the Seine, to its right (north) and left (south) banks, and to the uplands of the northern and southern sectors, above described. As to the river loops or meanders, let it be understood that the normal relation of valley-side spur, flood-plain scroll, and river meander is remarkably well shown at and below Paris: the spurs slope gradually to their end and their down-valley side; the spur end and its down-valley side are adjoined by a flood-plain scroll; the river flows along the steeper up-valley slope of each spur, and around the base of the opposite amphitheatre in the valley side.

With these points in mind, it is easy to remember that Vincennes and its woods, the Bois de Vincennes, lie east of Paris, north of the confluence of the Seine and Marne; Belleville is on the chain of hills near the Butte Chaumont; the center of Paris lies on the right bank of the first northward bend below the confluence of the Seine

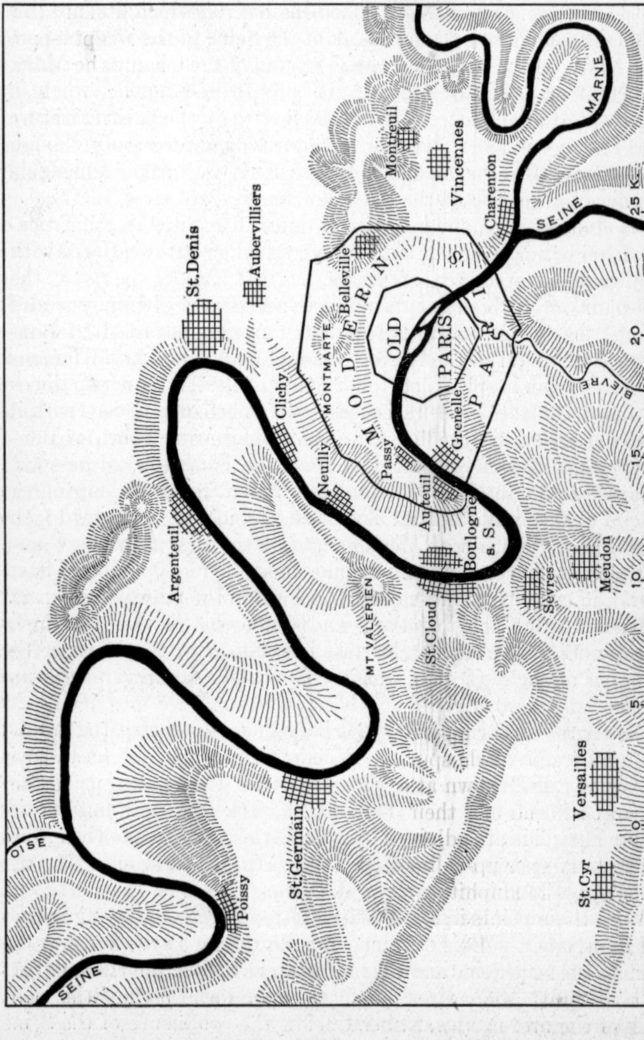
St.Denis
Aubervilliers
Montreuil
Vincennes
Charenton
MARNE
SEINE
Belleville
MONTMARTE
MODERN PARIS
OLD PARIS
Clichy
Neuilly
Passy
Grenelle
Auteuil
Boulogne s. S.
BIEVRE
Argenteuil
MT. VALERIEN
St.Cloud
Sèvres
Meudon
Versailles
St.Germain
Poissy
OISE
St.Cyr
0
5
10
15
20
25 K.

and the Marne; part of the flat flood-plain scroll which fits into this northward bend along the left bank of the Seine in the western part of the city is occupied by the parade ground of the Champs de Mars. Farther along the same bank is the suburb of Grenelle, where a famous artesian well, dependent for its flow upon the basin structure of the region, serves as an important source of water supply, though supplemented by other wells and more largely by surface aqueducts from neighboring valleys.

The steep slope of the opposite amphitheatre rises to the Trocadéro, from which a long spur declines gradually southwestward with Passy on its steep eastern side along the right bank of the river; the flood-plain scroll that wraps around the west side of this spur and fits into the first southward loop of the river is occupied by Boulogne-sur-Seine (57,027) with its Bois and by Neuilly, lately famous for its American hospital; part of the plain between these two towns is utilized for the level racecourse of Longchamps. Sèvres and St. Cloud are on the left bank in the amphitheatre south of Boulogne; the famous fortress of Mt. Valérien is on top of the spur, north of St. Cloud; St. Denis and Argenteuil mark the beginning and end of the amphitheatral slope around the first northward loop below Paris; St. Germain lies on the farther part of the next amphitheatre, which encloses the second southward loop, and its forest covers the spur which extends into the second northward loop; at the farther turn of this loop the Seine is joined by the Oise, on which Pontoise lies at the crossing of an ancient road. Poissy lies at the beginning of the next amphitheatre into which the third southern loop enters; and so on.

The degree to which the centralization of government, arts, industries, and traffic in Paris has been carried may be judged by the number of railways and national roads that converge to it, as the spokes of a wheel converge to the hub. A circle of 25 k. radius drawn around Paris cuts the main lines and various branches of the five chief railway systems of the West, the North, the East, the P. L. M. (Paris-Lyons-Mediterranean), and Orléans at some 15 points; it cuts also a similar number of national roads directed to as many important cities, some of which, like Bordeaux to the southwest, Lyons to the southeast, and Bâle in northern Switzerland, are 400 or 500 k. distant; short stretches of the roads to Lyons, Bâle, and Nancy are shown in Figs. 8 and 9.

The following pages are planned to make a circuit of the successive parts of the Paris basin, beginning on the east, and passing around by the north to the northwest, thus including as much of France as lies northeast of the Marne and the

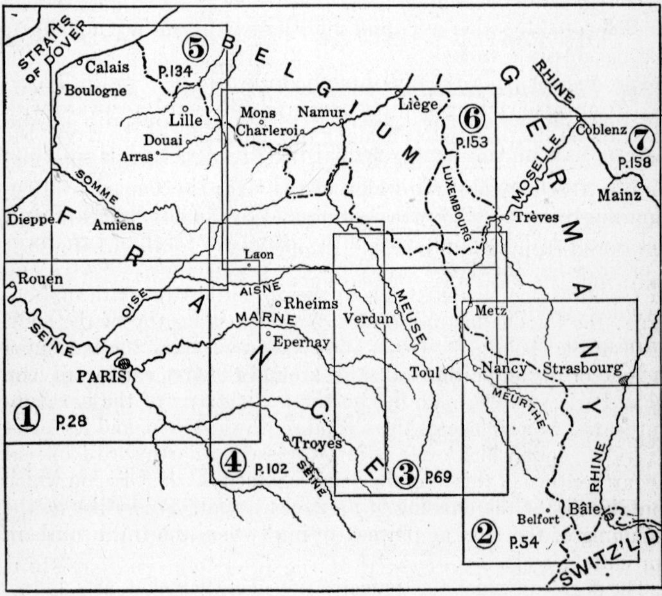

FIG. 12. INDEX OF OUTLINE MAPS

lower Seine. Brief accounts of adjacent regions farther northeast are added. The location of the outline maps which represent the areas described is indicated in Fig. 12; the maps will be referred to in the order here given from 1 to 7; the page numbers adjoining the map numbers indicate where the large-scale maps can be found.

CHAPTER IV

THE EASTERN HALF OF THE PARIS BASIN

20. *The Belted Relief of Northeastern France.* The concentric belts in which the stratified formations outcrop in the eastern half of the Paris basin from the Central Highlands on the south to the Ardennes on the north, and the correspondingly concentric arrangement of uplands (shaded and numbered from 1 to 8 in Fig. 13) and lower lands (unshaded) following the outcrops of stronger and weaker strata, like aureoles around the Paris center, are the leading physiographic features of northeastern France. There are indeed no better illustrations of this kind of topography in the known world. As the variety of features is large, their description cannot be short.

Each belt of weak strata is worn down, as in Fig. 14, in an unsymmetrical depression, limited on the east by the long and gentle ascent of the next underlying resistant strata, and on the west by the steep scarp of the next overlying resistant strata. Conversely, each belt of resistant strata remains in relief as an unsymmetrical upland ridge, having a broad crest with a steep escarpment or scarp descending eastward to the depression excavated on the next underlying weak strata, and a long gradual slope descending westward to the depression excavated on the next overlying weak strata. Hence the "grain" of the relief in this half of the Paris basin follows a series of upland belts arranged in concentric arcs, convex eastward, as above noted.

The eight upland belts are by no means alike. The first is of strong relief; the second is lower and of more delicate form; the third is the shortest of the series; the fourth and fifth are long and strong; the sixth is less distinct; the seventh is the least developed of all, and is characterized more by its limestone soils than by an unsymmetrical ridge form; the eighth is strong along the western border of the Vosges, but elsewhere is marked chiefly by its infertile sandy soils. The intermediate depressions also vary in dimension and in form: it should be understood that, although unshaded in Fig. 13, they are not always lowlands of small altitude above sea level, nor are they always plains; some of them have a rolling or hilly surface 100 m. or more in altitude; but they are always lower than the adjoining upland belts.

The resistant strata which determine an upland belt frequently crop out in bare ledges, or are covered only by a thin stony soil in the upper part of the steep scarp and over the broad crest of the upland; the weak strata, on the other hand, which are worn down in the longitudinal depression between two upland belts are usually concealed under a deep soil; the same is true of the lower slope beneath the steep scarp on one side of an upland belt and on the long descent of the other side.

Some of the upland belts are called *côtes* by the French, but the word has no special meaning; the Spanish etymological equivalent *cuesta* (pronounced in two syllables, *kwes-ta*), locally employed for certain unsymmetrical upland ridges on our Mexican frontier, has been adopted by a number of American geographers as a special name for forms of this kind and will here be sometimes used. The steep scarp of a cuesta is often called its face; and the long slope of the other side is called its back. The unsymmetrical depression or lowland between two cuestas has no technical name; it can hardly be called a "valley," because it is drained by many streams instead of by one.

While the upland belts are thus arranged in concentric arcs, the valleys of the larger rivers are otherwise disposed

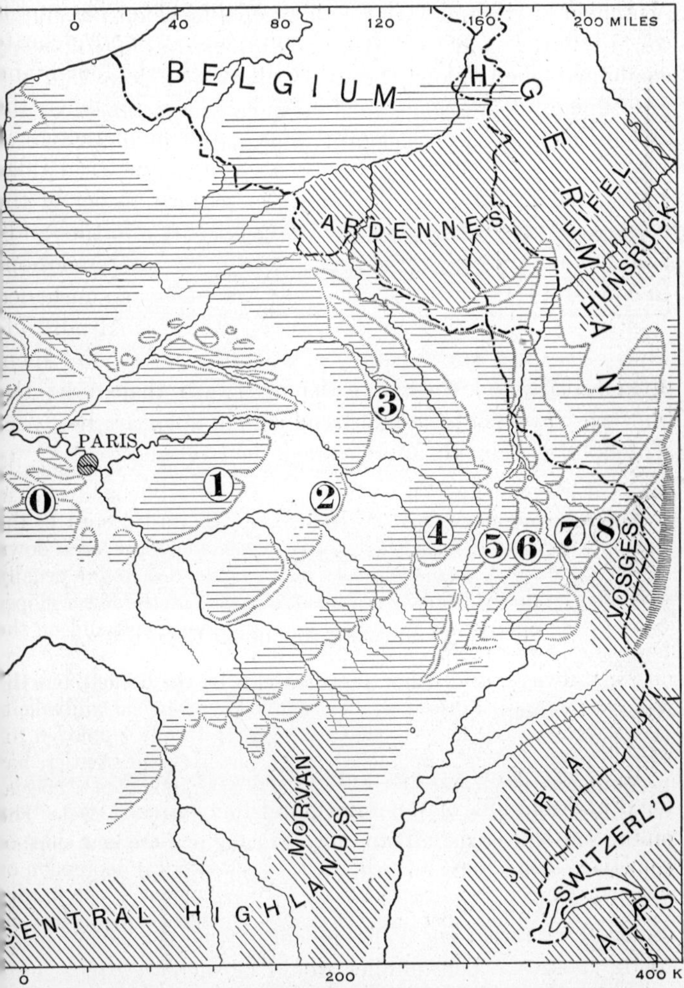

FIG. 13. PHYSIOGRAPHIC FEATURES OF NORTHEASTERN FRANCE

A number of rivers flow in roughly radial fashion, converging toward Paris; and as travel and transportation are today conducted chiefly along the river valleys, it follows that the inter-upland depressions are of secondary importance as lines of movement. Nevertheless it is primarily in terms of the

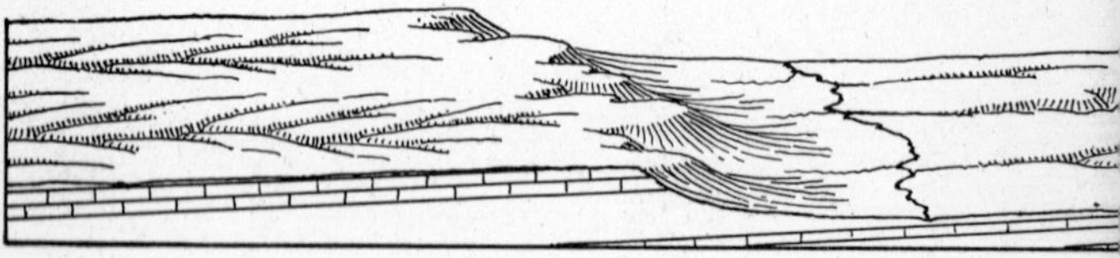

FIG. 14. UPLAND BELTS

upland belts and the depressions between them that the physical features of the half-basin east of Paris are best presented; after they are apprehended, the river and valleys, the

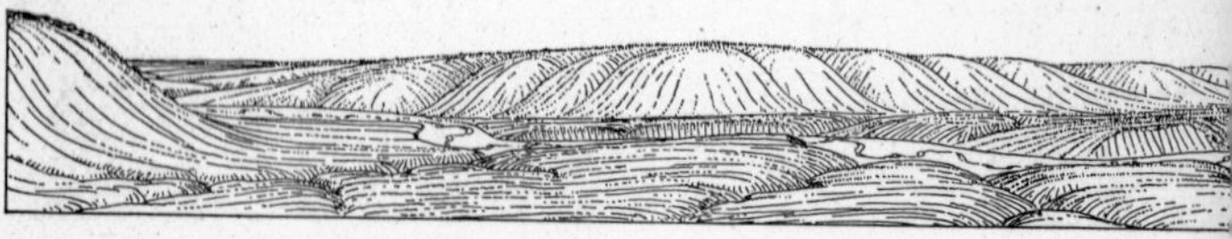

FIG. 15. THE STEEP SCARP OF AN UPLAND BELT

cities, railways, and other features can be duly located with respect to them.

21. *The Upland Belts as Natural Defences.* Without including the tabular hills of uppermost sandstones already described on the south of Paris, all but one of the eight more or less continuous upland belts, separated by longitudinal depressions and varying greatly in height, breadth, and pattern, are crossed by a radial line 350 k. in length from the basin center to the Vosges. The number would be increased if several subordinate cuesta-like belts or benches were counted.

It has long been remarked that this arrangement of the relief, which presents gentle slopes and broad crests for occupation by the home forces, and steep scarps to invaders from the east, provides a series of natural defences against an attack from the German frontier; but the defences were

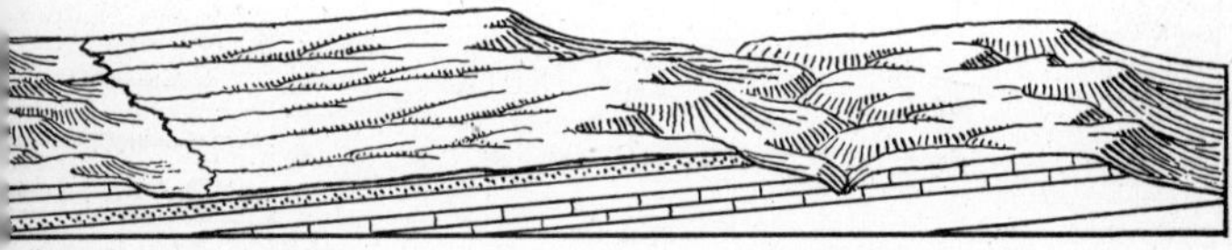

mediate Depressions

overcome in 1870 by the superior organization and prompt mobilization of the Germans, who advanced rapidly across the Lorraine saddle and over the upland belts to Paris. Since

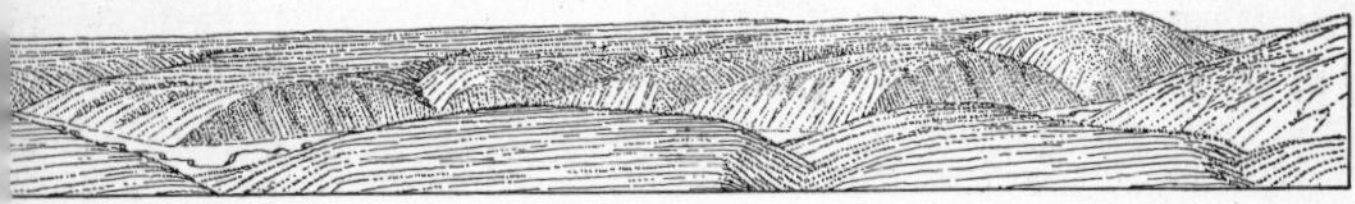

Fig. 16. The Long Back Slope of an Upland Belt

then the natural defences of the northeastern frontier have been reënforced by the construction of the chain of fortresses above-named — Belfort, Epinal, Toul, Verdun, Mézières, Maubeuge — from the Vosges to the Ardennes. Hence in 1914 the Germans, instead of again advancing from Lorraine, passed around the Ardennes, thereby deliberately violating the neutrality of Belgium, in order to attack France from the north, where most of the upland belts are wanting. There the defences, both natural and artificial, are weaker, the mining and industrial districts of the northern border were rich prizes, and the advance southward into the region north and east of Paris was along the grain of the relief rather than across it. It is for this reason that the three more southern fortresses of Toul, Epinal, and Belfort have not been attacked in the present war, that the northern (unfinished) fortresses of Maubeuge and Mézières were overwhelmed, and that the intermediate fortress of Verdun, where the fighting front crosses

FIG. 17. RELATION OF STRUCTURE AND FORM IN UPLAND BELTS

the line of fortresses, has been the scene of a prolonged struggle.

22. *Varied Features of Different Upland Belts.* Before proceeding to the description of the individual features in the eastern half of the Paris basin and its extension over the saddles of Langres and Lorraine, a brief statement may be made of the various ways in which the upland belts or cuestas depart from the simple forms of Figs. 15 and 16. The stronger and thicker the cuesta-making strata, the higher and bolder is the cuesta, as on the left in Fig. 14. A thin cuesta-maker between heavy bodies of weaker strata will make a low cuesta or upland, as in the middle of the figure. The weaker and thicker the intermediate strata, the broader, lower, and smoother is

he inter-cuesta depression, as to the left of the middle. If a thin ody of weak strata lies between two strong cuesta-makers, the esulting depression, as near the right end of the figure, will be nar-ow and shallow, so that it as well as the upland belts will be trenched y stream valleys; upland belts thus related may be described as verlapping, in contrast to the belts in the middle and near the left nd of the figure, which may be described as wide-spaced.

The greater the slant of the strata, the more direct the front of the pland belt, until when the strata are steep or vertical, as in the fore-round of Fig. 17, the belt becomes a narrow ridge: on the other and, the less the slant of the determining strata, the more irregular he upland front, until when the strata lie horizontal as in the back-round of Fig. 17, the upland becomes a table-land with an irregular nargin and detached outliers, as is the case in the first upland belt 00 k. northeast of Paris. Variations in the resistance, the thickness, r the slant of strata along their belt of outcrop therefore cause cor-esponding variations in the form of the upland belts.

The relation between river courses and upland belts is very aried. The simplest relation is that of a squarely transverse river, lowing with the slant of the cuesta-making strata, as in Figs. 15 nd 16; and such is the course of the historic Marne where it cuts narrow gateway through the strong upland next east of Paris, the irst of the series counting eastward to the Vosges, as will be again old in section 47. Gateways of this kind are of great importance s lines of travel: they can be strongly defended against invasion by ortifying the adjoining uplands. The Meuse, on the other hand, uts off a long segment from the curved front of the fourth belt, omewhat as in Fig. 18; the river enters obliquely from the south, uns behind the upland front as a chord passes behind its arc, and lows out obliquely to the north, as will be more fully described in ection 37. Many of the uplands are much dissected by the ravines f small streams, so that they form belts of hills, rather than a ontinuous upland surface.

Streams occasionally flow through an upland belt against the slant f the determining strata; several such streams cut trenches hrough the strong rampart of the fifth upland belt west of the Moselle as shown on the map, p. 69, and thus bring to that river he drainage of the broad lowland, known as the *Woëvre*, which there lies between the fifth and the fourth upland belts: one such stream heads at a point where the fourth upland is exceptionally narrow and it is there that the Germans, advancing from Metz, pushed forward their front in a salient which overlaps the scarp of the fourth upland and includes St. Mihiel on the chord-valley of the Meuse as will be stated more in detail in section 39. If the details presented here and on the following pages seem complicated, let it be

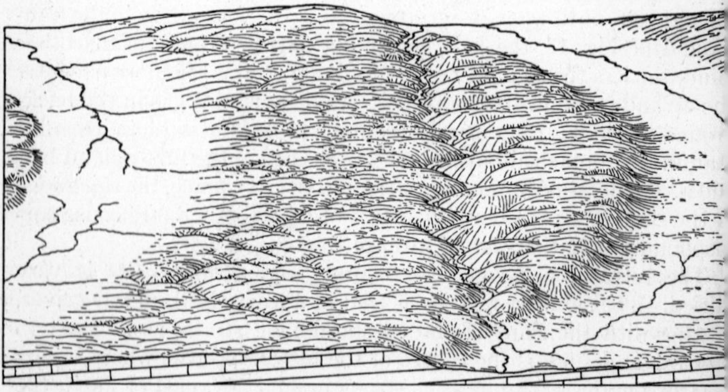

FIG. 18. AN UPLAND BELT WITH A DETACHED SEGMENT

remembered that they are much simpler than the varied forms of nature.

It may be noted that the upland belts or cuestas are not so easily recognized on a large-scale map as on the ground, for the downward view of a map shows the many valleys by which an upland and the adjoining depressions may be dissected, and therefore does not set forth the unity of the upland nearly so well as when it is seen in outdoor nature. But outdoor views of the uplands may also fail to reveal their true character if they are seen from the floors of the larger transverse valleys that are followed by the chief routes of travel; for if, as is often the case, the intermediate depressions as well as the uplands are trenched by the transverse valley that an observer follows, an upland seen from the bottom of such a valley may be regarded simply as a hill somewhat higher than its neighbors and the long continuity of its relief will not be recognized. It is in the almost horizontal views from hilltops, whence all the valleys but those in the foreground disappear, as in Fig. 15, that the continuity of an upland belt is best perceived, especially if its front rises only 50 or 60 m. above an adjoining dissected depression.

It would greatly facilitate the description of the upland belts if each one had a name for itself, but even the strongest of them vary so much in form and are cut across by so many valleys that their continuity has never been recognized in popular nomenclature. The habit followed by some geographers of designating the belts by the names of the geological formations to which their strata belong — as bathonien, bajocien, kimmeridgien, etc. — is unsatisfactory, because such names are technical and unfamiliar, and because attention is thereby turned too much from their exterior form to their interior constitution. They might be named after the cities that lie upon or in front of them; thus the fifth could be called the Langres-Nancy-Metz cuesta; but such compound names are not convenient for frequent use. The device of numbering the belts eastward from Paris, as in Fig. 13, has at least the merit of simplicity and of giving easy indication of their relative positions, but the numbers thus employed are not in current use in France.

One reason for the lack of simple geographical names for the upland belts is that many of them are so long that they pass from one ancient historical province or modern political division to another; and as the leading French students of geography have entered the subject from the historical rather than from the physiographic side, even they have not yet introduced generally accepted names for these striking features. Partial exception to this statement may be made for the third and shortest belt, which lies chiefly in the western part of the district of Argonne and which, being forested for much of its length, is known as the Forest of Argonne; but this name is not applied to its northwesternmost extension.

CHAPTER V

THE VOSGES AND THE ADJOINING REGIONS

23. *The Highlands of the Vosges.* The eastern side of the Paris basin is limited by the Vosges (German, *Vogesen*, or *Wasgau Gebirge*), a mountainous highland which should be considered, as shown in Fig. 19, in association with the similar highland of the Black Forest (German, *Schwarzwald*), 45 k. farther east, beyond which a covering series of stratified formations slopes gently eastward toward the basin of the upper Danube and thus roughly corresponds to the series of covering

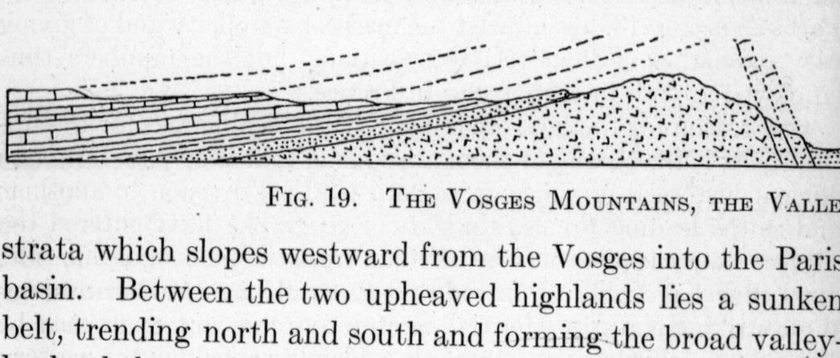

FIG. 19. THE VOSGES MOUNTAINS, THE VALLE

strata which slopes westward from the Vosges into the Paris basin. Between the two upheaved highlands lies a sunken belt, trending north and south and forming the broad valley-lowland that is followed by the middle Rhine in the stretch from its narrow passage by Bâle (German, *Basel*) between the Black Forest and the Jura mountains on the south, to its entrance into the narrow gorge through the Slate mountains, 280 k. to the north. In consequence of this structural arrangement, both of the upheaved highlands have steep slopes toward the intermediate sunken area, and more gradual slopes toward the basins of overlapping strata.

The Vosges proper, consisting in great part of resistant crystalline rocks, increase in breadth from 30 k. in the north to 50 k. in the south, and measure about 110 k. in length along their north–south crest, where their height ranges from 1000 to 1400 m. The highest point is the Ballon de Guebwiller, 1426 m. in altitude, near the southern end of the highland. The summits are usually rounded, dome-like masses, covered with forests for the most part, though the highest domes are treeless above 1300 m.; but many valley heads, especially on the eastern slope, are steep and craggy. The broad crest of the highland is so little notched that no railroads cross it, though branch lines enter the valleys on either side.

The crest of the Vosges is however traversed by several roads, which like the railroads follow the valley bottoms into the mountains as far as the ascent is not too steep, but which on reaching the

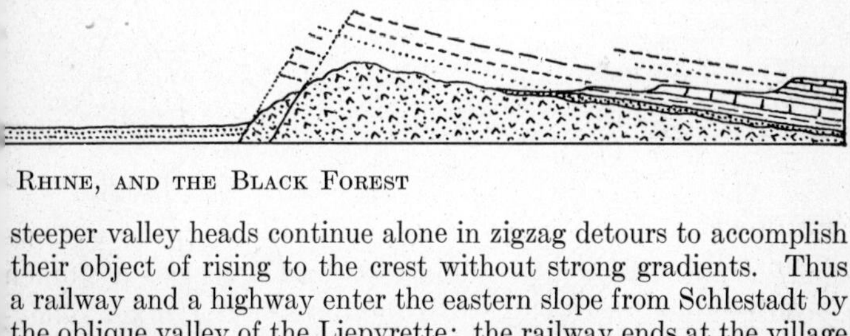

RHINE, AND THE BLACK FOREST

steeper valley heads continue alone in zigzag detours to accomplish their object of rising to the crest without strong gradients. Thus a railway and a highway enter the eastern slope from Schlestadt by the oblique valley of the Liepvrette; the railway ends at the village of Ste. Marie aux Mines (380 m.?; German, *Markirch*); the highway continues in zigzags to a pass (780 m.) near the mid-length of the range, and descends similarly to the upper valley of the Meurthe where it meets a railway at St. Dié (356 m.); the direct distance between the two railways is 16 k. Similarly a railway and a highway ascend from Molsheim by the oblique valley of the Bruche in the northern Vosges to the valley head; there the highway continues over a notch and descends to a railway on a headwater branch of the Meurthe northeast of St. Dié, but on the French slope

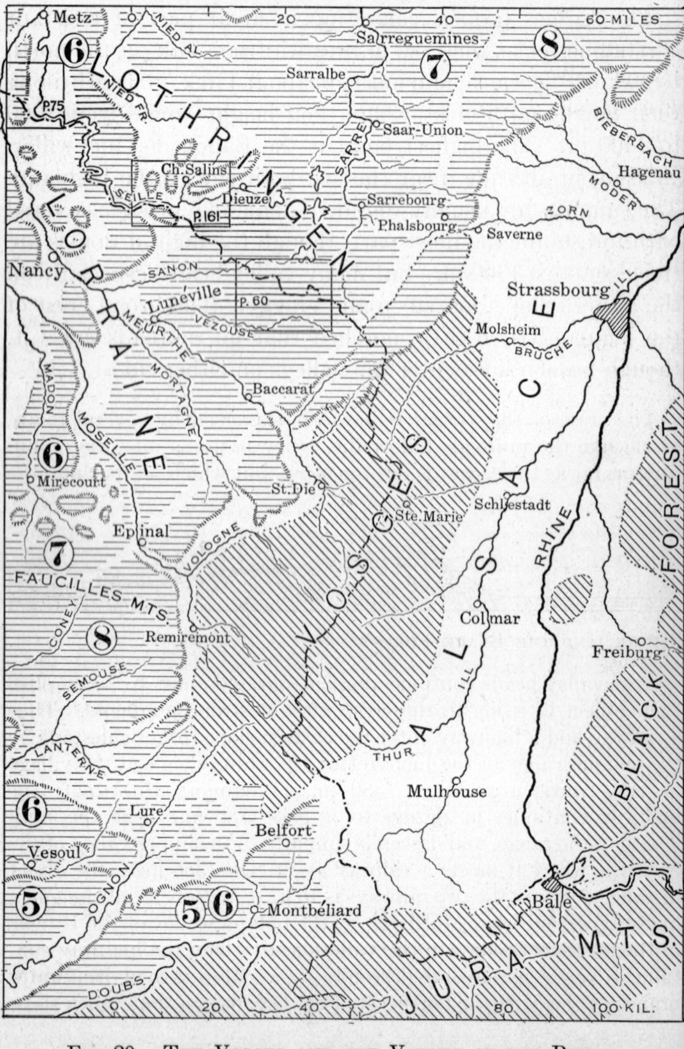

FIG. 20. THE VOSGES AND THE VALLEY OF THE RHINE

the road is not so well constructed. Again in the southern Vosges, a railway and a highway from Mulhouse ascend the valley of the Thur; the highway zigzags over a pass and meets a railway on a head branch of the Moselle. Next to the southwest one of the finest of the mountain roads lies altogether in French territory as it passes near an angle of the frontier from a railway at the head of the Moselle southward over a shoulder of the Ballon d'Alsace (1244 m.) near the southern end of the range, and, descending, joins a railway leading to the fortress of Belfort: its repeated zigzags lace across a convex spur on the north side of the pass, and around a concave ravine head on the south side.

The steep eastern slope of the Vosges, deeply dissected by many narrow valleys and ravines, is heavily forested. Except near the southern end of the range, the larger valleys trend obliquely northeastward; the spurs between them are elaborately carved by side ravines. A knob (German, *Kopf*) at the end of a spur (936 m.) on the eastern slope of the Vosges near their southern end and 20 k. east of the crest line, named Hartmannswillerkopf ("the knob of Hartmann's hamlet") from a village at its base, has become notorious during the present war from being occupied by the French; it surmounts the plain by over 600 m. The broad lowland of the middle Rhine, east of the highland slope, measures 40 k. in width; near the highland base it is occupied by rolling hills of small relief at an altitude of 200 or 300 m., and farther east by a broad and flat river plain 150 or 200 m. in altitude. The western part of the plain 10 k. east of the highland base is drained northward by the Ill, which rises in the Belfort depression south of the Vosges; 15 k. east of the Ill is the Rhine, which formerly flowed in many tangled or braided channels, but which is now artificially restrained to a single channel of gentle curvature through much of its course: near the junction of the two sub-parallel rivers east of the northern end of the Vosges lies the famous city of Strasbourg (German, *Strassburg;* 178,891.)

24. *Alsace.* The present boundary between France and Germany lies, as above noted, along the crest of the Vosges. All the eastern slope of the highland and the plain below it as far as the Rhine, from the Swiss boundary at Bâle past

Mulhouse (German, *Mülhausen;* 95,041) and Colmar and beyond Strasbourg, formerly constituted the French province of Alsace. Since 1871 the province has been German territory under the name of Elsass. The people here habitually speak two languages, and most places have two names, one French, one German. In 1872, 45,000 of the inhabitants withdrew into France in order to avoid becoming German subjects: many more French citizens, equally loyal yet unable to move away, were constrained to change their nationality; but they are "*français quand même!*"

25. *The Uplands West of the Vosges.* West of the Vosges crest line, the mountainous area has its greatest extension between the upper Moselle and Meurthe, where the ridges (750–950 m.), mostly forested, are divided by a labyrinth of irregularly branching valleys, from 200 to 400 m. in depth. The main valleys of the Moselle and the Meurthe follow generally northwestward courses; their floors have a width of one or two kilometers; their descent is more gradual than that of the narrow, eastern valleys. Two lakes (Gérardmer, 631 m., Longemer, 716 m.) occur in branch valleys of the upper Moselle. St. Dié lies on the Meurthe within the margin of the mountains; the strongly fortified city of Epinal is at the mountain margin 40 k. farther west on the Moselle. Beyond the mountains the relief diminishes to a more moderate measure as the crystalline rocks are irregularly overlapped by the lowest strata, the eighth and seventh belt-makers, of the Paris-basin series.

The district which extends west of the Vosges to the sixth upland belt is occupied by the lowest sandstones and limestones of the Paris-basin series (Fig. 21). The basal member or eighth belt-maker is a resistant sandstone, which weathers to an infertile soil and is therefore generally forested. It is frequently strong enough to rise in high uplands or ridges (550–800 m.) along the mountain margin. It is followed on the northwest by the rolling uplands of the broad seventh belt (275–350 m.), composed chiefly of limestone strata which produce a fertile soil and are therefore generally cleared and cultivated. These strata are usually without distinct topographic expression as a broad cuesta or unsymmetrical ridge; but in certain areas their edge determines a well defined east-facing scarp. Their rolling surface declines gently westward, and at distance of some 40 k. from the Vosges, they are overlapped by stronger limestones which rise in a well

FIG. 21. THE SIXTH, SEVENTH, AND EIGHTH BELTS WEST OF THE VOSGES

marked scarp to the higher surface (350–400 m.) of the sixth upland belt, sometimes forested but more generally cleared.

The general trend of the eighth and seventh belts hereabouts is from southwest to northeast: thus they enter well into France west of the southern Vosges, where they adjoin the saddle of the Langres plateau, as shown on the map, p. 69; while to the northeast of the frontier they extend far beyond the northern Vosges into Germany, where they form the Lorraine plateau and the Hardt, to be described in Chapter XIII. Within French territory, the two belts are cut almost squarely across by the valleys of the Meurthe and the Moselle, and they will therefore be described in three sections: the first, about 20 k. wide, from the frontier to the Meurthe valley; the second, about 30 k. wide, between the two valleys; the third, extending 60 k. southwest of the Moselle valley.

In the section to the northeast of the Meurthe, the basal sandstones of the eighth belt are locally doubled, as above figured, in two mountainous, forested ridges. The first ridge, from 600 to 800 m. in altitude, is separated from the irregularly subdivided hills and mountains of the crystalline rocks on the southeast by the valley of the Rabodeau, and from the second ridge, which seldom exceeds 700 m. in altitude, by the valley of the Plaine river; both of these streams flow southwest to the Meurthe. To the northeast, beyond the frontier, the high sandstone hills flanking the Vosges are irregularly dissected by the headwaters of the Sarre. The sandstones are followed on the northwest by the overlying limestones, which form a lower rolling surface subdivided by the valleys of the Vézouse, the Sanon, and their branches, partly shown on the 1:100,000 map, pages 60, 61.

When regarded northeastward from favorable points of view, the general profile of the cleared hills which make up the rolling surface between the main valleys may be seen to rise toward the mountains, thus forming the seventh upland belt, and then fall off in a low scarp before the ascent of the first sandstone ridge is begun: but the scarp is so often cut back by many little valleys and ravines that its scalloped front has little continuity: these features are continued northeastward, beyond the frontier. If the rolling surface is followed northwestward, residual hills of the next overlying strata, mostly forested, are found along the divide between the Sanon and the Seille near the frontier, between the Sanon and Vézouse, and less distinctly between the Vézouse and the Meurthe; then beyond the junction of the Sanon and the Meurthe a well defined scarp rises to the broad surface (350–400 m.) of the sixth upland belt, which will be followed from south to north in a later paragraph. The streams by which this section is limited and divided have the habit, like many others in this region, of flowing in a very irregular course on the flat floor of a winding valley, one or two k. wide; a reason for this behavior will be suggested in the account of the Meuse, in section 38.

It was in this district, between the Vosges on the southeast and the sixth upland belt on the northwest, that the German army in August, 1914, crossed the upper Meurthe and advanced half way over the hills to the upper Moselle; they were later forced back almost to the frontier, northeast of the Meurthe, where the fighting front has since remained with little change. Ruined villages, such as Vitrimont on the Meurthe below Lunéville and Gerbéviller on the Mortagne, mark the temporarily invaded area.

The chief city of this area is Lunéville, on the widened valley floor (230 m.) at the confluence of the Vézouse and the Meurthe. The Eastern railway, coming from Paris and Nancy, ascends the Meurthe valley to this point and then turns up the Vézouse toward the frontier. Branch lines run up other valleys to the Vosges, and up the Moselle to Epinal and beyond. The valleys are also followed by branching highways. An important canal, crossing the Meuse and the Moselle on its way from the Marne to the Rhine, turns from the valley of the Meurthe and ascends that of the Sanon to the frontier, the irregular course of which is described in section 27.

In the section between the Meurthe and the Moselle, which is unequally divided by the intermediate valley of the Mortagne, the basal sandstones, slanting gently northwest, assume the form of a well defined upland belt (700 m.) somewhat southeast of the line connecting St. Dié and Epinal, with a strong frontal scarp toward the Vosges; but instead of being continuous, the upland is irregularly incised by many narrow valleys, and the scarp is worn into a frayed-out pattern. These details of form are beautifully displayed in the elaborately dissected upland area west of St. Dié, which bears the Mortagne and other forests; yet complicated as the area is when seen in plan, all its parts are merely the dissevered elements of a cuesta-like upland; and the recognition of this fact greatly facilitates the appreciation of many details that might otherwise seem unrelated.

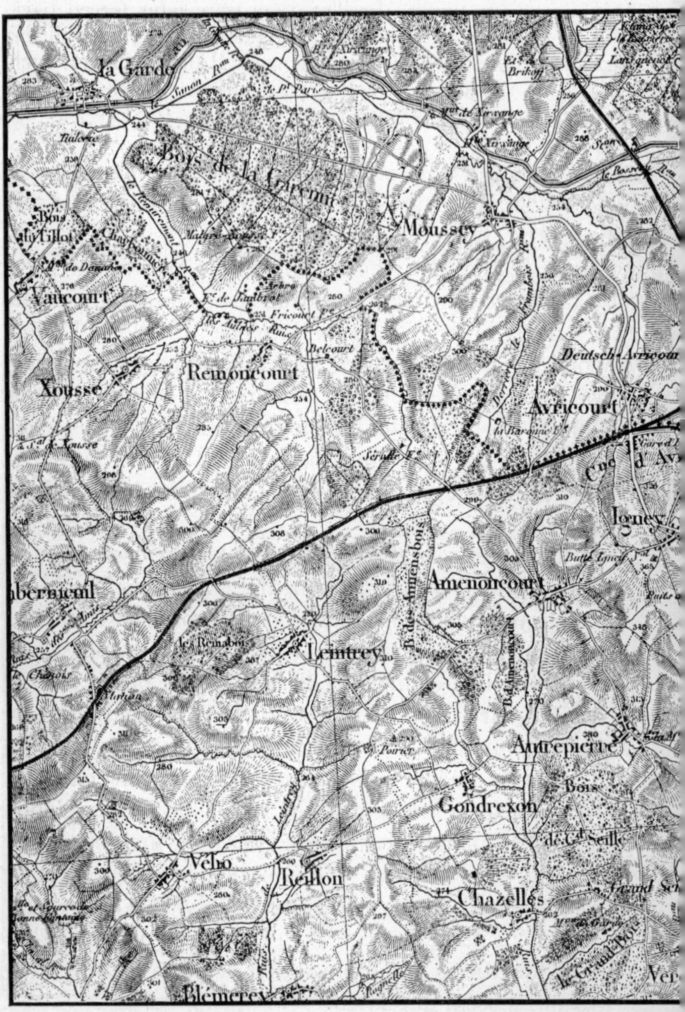

FIG. 22. THE FRONTIER ON

ENTH UPLAND BELT

The following limestone belt repeats the features above described, except that more of its uplands are forested. The tabular limestone surface of the sixth upland belt (350 m.) is well developed, though of moderate extent, in the narrowed space where the Meurthe and the Moselle approach each other. The chief place in this section is the fortified city of Epinal (340 m.), commanding the narrows of the elsewhere open Moselle valley where it cuts through the slanting uplands of the basal sandstones.

The third section, southwest of the Moselle, is much more extensive than the other two: it is characterized by the great extension of the basal sandstones which here lie nearly horizontal and stretch 30 k. southwestward in a tabular upland, cut into separate portions by the narrow valleys of the Coney, Semouse, and Lanterne, headwaters of the Saône. A noticeable feature of this section is a range of hills, known as the Monts Faucilles, which divide the side branches of the Moselle southwest of Epinal from the headwaters of the Saône; the hills are merely remnants of the next strata overlying the sandstones, which are naturally enough not yet completely worn away along the divide.

Southwest of the sandstone area the overlying limestones, the seventh belt counting eastward from Paris — see the map on p. 69 — rise in a well defined, east-facing scarp west of the uppermost Saône, and the upland surface beyond declines gently westward to the head branches of the Meuse; hence a cuesta-like upland is here formed with much more distinct expression than in the same limestone belt farther northeast.

26. *The Sixth Upland Belt.* The limestones of the sixth belt form a well-marked upland, 400 m. in altitude, with a distinct east-facing scarp, cut into many scallops by transverse streams. Near its southern beginning, it forms the divide between western branches of the Saône and the head of the Marne; from 20 to 40 k. farther north it is obliquely cut through by several branches of the Meuse which rise on the back slope of the seventh upland belt. The longitudinal depressions which adjoin the upland on either side are not

smooth surfaces, but are incised by the valleys of the oblique streams, though to a less depth than that of the valleys through the upland.

It was on the eastern slope of this upland south of the Meuse branches and on the Apance, a small branch of the upper Saône, that a super-Zeppelin, "L-49," was forced by five French aëroplanes to land during its return from a raid on England. It was capable of making 55 or 60 miles an hour, with a crew of 18 men, two machine guns, and two tons of bombs, and had reached a height of 4½ miles over London, where the temperature was — 33° C. with a strong north wind. Some of the men had their hands frozen and were half stupefied with the cold. They were prevented from destroying the airship after landing by a sportsman, who happened upon them; the airship was thus captured intact and carefully studied by French experts. On the same date the frame of a second airship, set on fire by French guns, fell at St. Clément on the Meurthe above Lunéville; and a third was destroyed by its crew after landing in the French Alps.

Beyond the Meuse branches, the upland belt turns northeastward and the upland is obliquely traversed by the north-flowing Madon, which joins the Moselle above Toul; at the entrance of this oblique valley lies Mirecourt: several detached outliers rise southeast of the upland scarp toward Epinal. After resuming its northward course, the upland scarp is skirted on the east for 15 k. by the Moselle; when this river approaches within 12 k. of the Meurthe, both rivers flow northwestward through oblique gateways in the upland belt. A fine view is obtained from the crest of the upland between the two rivers far southeastward, up the two valleys and across the uplands between them to the Vosges, 50 k. distant: in the opposite direction, the strong scarp of the fifth upland belt is seen beyond the depression that separates the two uplands. North of the Meurthe, the sixth upland belt advances northeastward and broadens in somewhat tabular form: there it is cut through by the Seille, a small stream of very winding course, which marks the frontier: the further northward extension of the upland will be described in chapter XIII.

It is noticeable that the rivers hereabouts do not follow the depression between the sixth and fifth uplands, but cross it and the adjoining uplands with little regard to the relief of the surface. On the other hand the depression, although by no means a level surface, is continuous enough to be followed by lines of communication and occupied by villages, both of which avoid the uplands. The river gateways or gaps through the upland belts open easy lines of travel between the neighboring depressions.

27. *The Frontier from the Sixth Upland Belt to the Vosges.* The frontier dividing Lothringen from Lorraine, as established in 1871, ascends the incised course of the meandering Seille southeastward through the sixth upland belt, but departs from that small stream on reaching the more open country of the seventh belt and runs across hills and valleys to the Vosges. After crossing the upper Sanon it runs, as in Fig. 22, about halfway between the headwaters of the Vézouse and the Sarre, traverses the double sandstone ridges at the head of the Plaine, and then rises to the mountain crest. Through the middle of this distance of 60 k., the boundary line is not signalized by any striking topographic features: the limestone country on one side of it is much like that on the other. The adjoining German area will be described in section 68.

It was across the open upland of the seventh belt and thence westward through gaps in the sixth belt within the 70 k. space between the fortresses of Epinal and Toul, that French military writers during the period following the war of 1870 thought the next German attack would be made, and truly enough a strong advance was there attempted, as above noted: but the main line of the German invasion in 1914 was through neutral Belgium.

CHAPTER VI

FROM THE PLATEAU OF LANGRES TO LORRAINE

28. *The Plateau of Langres.* The saddle where the strata of the Paris basin arch over the high depression between the still higher masses of the Vosges on the northeast and the Morvan on the southwest is known in part as the plateau of Langres, from an old fortified town that occupies one of its spurs: the saddle constitutes the northwestern half of the old province of *Bourgogne* or Burgundy. It is composed of resistant limestones that ascend gradually from the northwest, arch over the saddle at an altitude of 550 m., and descend southeastward in a rapid and well dissected slope, known as *la Côte d'Or*, famous for its " Burgundy " vineyards.

The slope ends in an escarpment, 100 or 150 m. in height, made ragged by the notches of many small streams, beyond which the broad lowland of the Saône basin — the plain of *la Bresse* — is outspread.

The southwestern part of the long and gradual ascent of the plateau is trenched and divided into several long strips by the subparallel headwaters of the Armançon, a branch of the Seine system which, more directly than any other, lies in the up-stream prolongation of the trunk river between Rouen and Paris. Here from time immemorial travel and traffic between the southern valley of the Rhone — "Provence" — and northern France have crossed the plateau: a Roman highway, placed on one of the narrow plateau strips for safety from attack, is still traceable for 25 k.; a detached hill now known as Mt. Auxois (418 m.) at the end of one of the strips is the site of Alesia (" Ipsum erat oppidum Alesia in colle summo, admodum edito loco, ut nisi obsidione expugnavi non posse videretur "), a stronghold where Vercingetorix, who had gathered all the tribes of Gaul to his aid, was finally overcome by Caesar, B.C. 52. The ancient name is still preserved in Alise Ste. Reine, a village at the foot of the hill. The plateau is now crossed by a national road, a canal connecting the Seine and Rhone river systems, and the main line of the "P. L. M." railway which, after ascending the Armançon valley near each other, follow different lines over the high ground, but which all come together again at the base of the Côte d'Or, where the city of Dijon (76,847) lies at a valley mouth.

29. *The Fifth Upland Belt: Southern Part.* The fifth upland belt, maintained by a series of resistant limestones and shown in most of its curved length on the map, p. 69, is one of the most remarkable members of the upland series. Its southern part is divided into two steps, of which the eastern and lower one forms a bench below the higher or main member: but the two parts approach and coalesce farther north. Near the southern beginning of this complex belt, the crest of the main upland is sharply defined with a precipitous but irregular scarp, to the east of which the lower member or bench, the sixth upland belt and the seventh follow in regular order, but with very irregular pattern when seen in plan, as shown in Fig. 23.

When it is remembered that access from Germany to the plain (*la Bresse*) of the Saône through the narrow and hilly gateway between the Vosges and the Jura is guarded by the strong fortress of Belfort, and that advance from the plain of the Saône over the plateau of Langres involves the ascent of the steep Côte d'Or or the traverse of the four benched uplands to the north of it, before the basin of Paris can be entered, it is clear that invasion of France from this side can not be easily accomplished.

The ancient walled city of Langres, from which the adjoining plateau takes its name, stands on a sharp spur-end in the main scarp (470 m.) of the fifth belt, where it enjoys a broad eastward view across the bench and the lower belts, and commands the pass (400 m.)

between the headwaters of the Marne and the Saône. This city is one of the few in France situated on the crest of an upland belt; its importance is less now than formerly, for modern travel and traffic by road, canal, and railway pay little heed to the old-fashioned town, perched above the cliffs over the pass; its isolation combined domination with safety in the middle ages, but now turns to its disadvantage.

The northwestern slope of the main member of the fifth belt is a forested and sparsely inhabited upland for some distance north of Langres, almost waterless because its determining limestones are pervious to rainfall: it is deeply incised by many northwestward

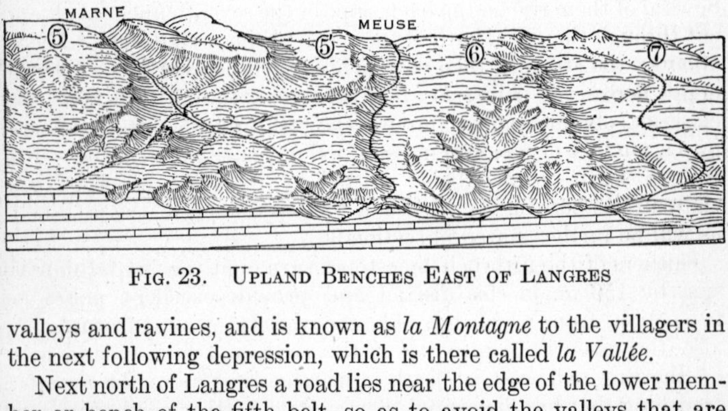

FIG. 23. UPLAND BENCHES EAST OF LANGRES

valleys and ravines, and is known as *la Montagne* to the villagers in the next following depression, which is there called *la Vallée*.

Next north of Langres a road lies near the edge of the lower member or bench of the fifth belt, so as to avoid the valleys that are incised in its back slope by the headwaters of the Marne. This winding river, which here pursues a northwestward course, cuts a valley gateway through the main body of the upland; the valley sides decrease in height as the river winds its way through the lowering back slope of the upland. A similar through-valley is cut by the Rognon, a branch of the Marne, which competes with the head of the Meuse for the drainage of the sixth upland hereabouts.

30. *The Fifth Upland Belt: Middle Part.* Beyond the Marne and 30 k. north of Langres, the two members of the fifth upland belt, which were 5 or 10 k. apart and of very irreg-

ular pattern near Langres, approach to within 2 k. of each other, and the general line of their double scarp becomes fairly direct: thus the belt assumes a simpler form. The sharp edge of the main scarp rises above the bench in a strong crest (480 m.), which retreats where valleys are cut through it and advances between them in pointed salients; but all its parts are systematically related as the dissevered members of a single upland belt.

Between 40 and 70 k. north of the Marne breach by Langres, five neighboring eastern branches of the Meuse deserve mention. Several of them rise on the back slope of the seventh upland belt and cut through the sixth, both of these belts being well developed hereabouts; the streams, then entrenching their courses to a moderate depth across the depression between the sixth and fifth belts, cut close-set, oblique valleys through the fifth upland, thus dividing it into a number of separate masses (480 m.) of small area, the southeastern scarp of each being skirted by its lower bench. Neufchâteau lies here on the Meuse in the back slope of the fifth belt, as will be further stated in section 36.

Outliers of the fifth-belt limestones surmount the frontal depression by 150 m. in this district and provide excellent points of inspection, from which not only the variety of local features but also the essential continuity of the sixth and fifth upland belts on the east and west is clearly apparent, in spite of the repeated trenching that has dissevered their parts. Farther north, the scarp of the fifth belt is deeply indented by frontal ravines, but the belt as a whole is not cut through in the 30 k. stretch beyond Neufchâteau.

The Moselle cuts an important breach through the fifth upland belt, and 12 k. farther north the Meurthe enters obliquely into the upland; but this river, instead of flowing through the upland, merely cuts off a segment of its front, as is further stated below. The broad top (400 m.) of the upland north of the Moselle breach is covered with the Forest of Haye; its bold east-facing scarp, sharply indented by many

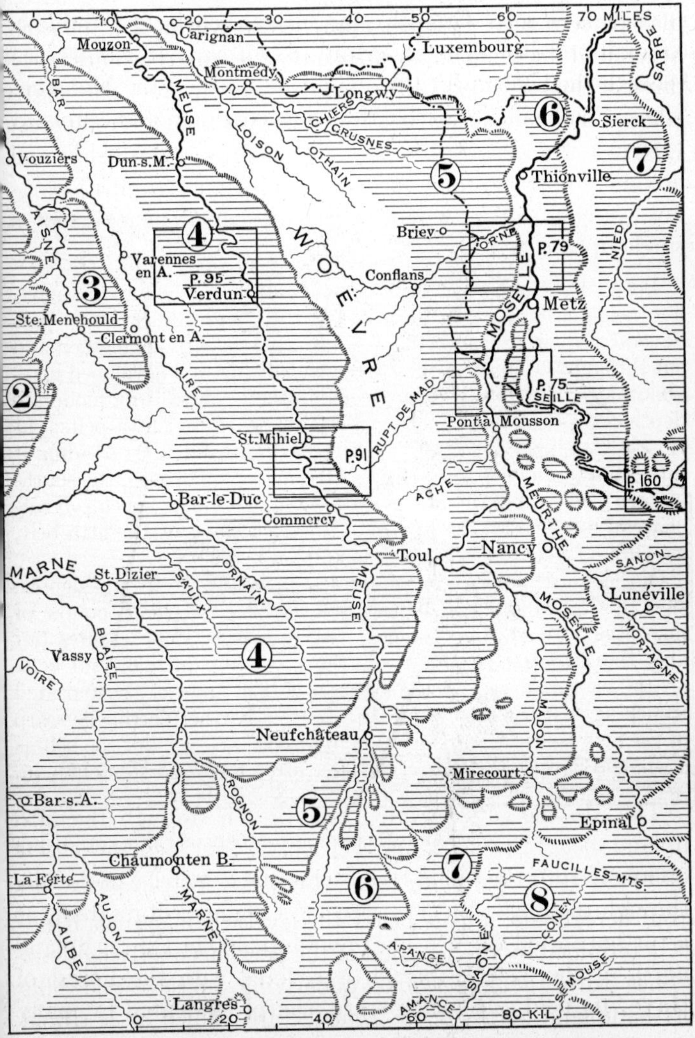

FIG. 24. THE FOURTH, FIFTH, AND SIXTH UPLAND BELTS

ravines, is closely skirted by the frontal bench, here reduced to small breadth. The important city of Nancy (220 m.) lies beneath the bench near the oblique entrance of the Meurthe. Many small villages lie on the benched slope of the upland

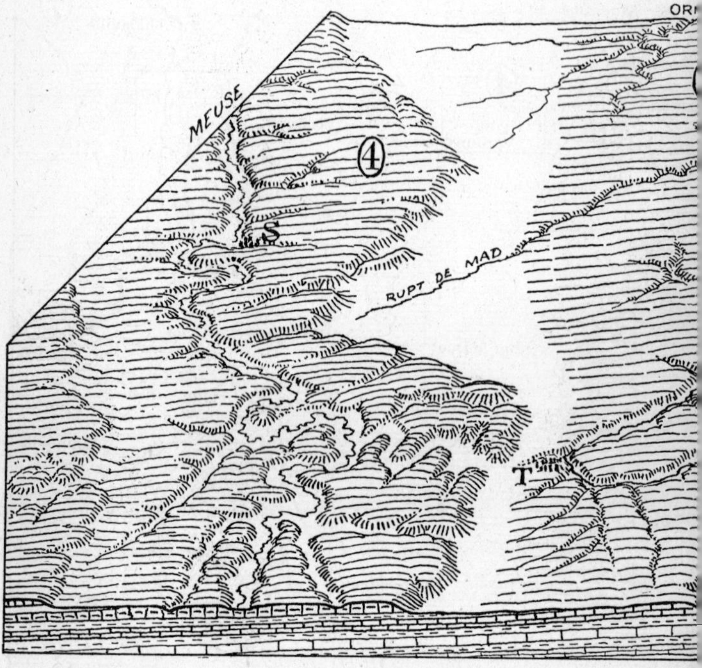

FIG. 25. THE MEUSE, THE MOSELLE, AND THE MEURT

Abbreviations: L, Lunéville; M, Metz; N, Nancy; S, St. Mihiel; T, Toul.

front hereabouts, a little below the level where springs issue, as determined by the junction of the pervious overlying limestones with the impervious underlying marls. The hills of the upland north of the Meurthe are known as *le Grand*

Couronné; they were held by the French against assaults directed against Nancy in the second month of the war, when large numbers of Germans — 40,000, it is said — were killed at the base of the hill slopes. Although unsuccessful in their

THE FOURTH, FIFTH, AND SIXTH UPLAND BELTS

main object, the enemy front has been maintained but a short distance away, and Nancy has been subject to bombardment ever since.

The repeated transverse dissection of the upland belt gives it an irregular, indeed a confused outline on the map, because the stream courses are not geometrical in their arrangement; but if the separated parts of the upland are carefully examined the systematic control of their unsymmetrical, east–west profiles by the moderate slant of their strata becomes apparent, as is shown in the bird's-eye diagram, pp. 70, 71. Each part of the upland resembles a slanting bastion, highest at the apex of its advancing and scalloped scarp, regularly declining along its indented flanks; the successive bastions are of different size and pattern in plan, according to the arrangement of the transverse streams; but in profile they are all much alike. The irregularly detached hills beyond the Meurthe are of more tabular form, having rather steep slopes on all sides.

31. *The Elbow of the Moselle.* The course of the main rivers in this district is peculiar: the Moselle formerly flowed northwestward not only through the sixth and the fifth upland belts, as it does today, but into the fourth also, where it joined the Meuse. At that time as now the north-flowing Meurthe entered into the fifth upland belt, but instead of running through it, followed a deeply incised valley a few kilometers west of the upland front and then ran out again. A much dissected segment of the upland, 40 k. in length, was thus cut off from the main mass. Farther on, the river, leaving the upland intact, pursued its course northward and northeastward to the Rhine. Where the two rivers entered the fifth belt, the Meurthe was the more deeply incised in the frontal depression by some 50 m.; and as a result one of its branches, heading to the southwest in the depression between the fifth and fourth uplands, slowly extended its length by retrogressive erosion in the weak strata there occurring and thus in time tapped the Moselle, diverted it from the Meuse which was thereby much diminished in volume, and led it to join the Meurthe which was thereby as much increased.

At the point of diversion, the Moselle, now narrowly entrenched below the broad floor of its former course, makes a sharp turn, known as an " elbow of capture "; there stands the fortified city of Toul. The former winding course of the Moselle, known as *le Val de l'Ane*, through the fourth upland belt will be described below. These changes are all long prehistoric: had they happened during the course of human history, a single name, such as Meurthe, would undoubtedly have been applied to the whole length of the original tributary of the Rhine, and another name, such as Moselle, would have been given to the shorter river above the point of confluence to which it was led by capture. Unfortunately, rivers have not been named in view of their origin and evolution, but in an arbitrary and often unreasonable fashion: hence the name of the Moselle like its flowing current is now continued to the Rhine; and the name, Meurthe, is limited to the upper part of the aboriginal river.

32. *The Fifth Upland Belt: Northern Part.* The segment of the fifth upland belt, set off as above noted from the main body by the valley of the Meurthe-Moselle, is widest at the southern end, where it is cut into several irregular tabular masses (400 m.), which advance eastward and surmount by 150 m. the gentle back slope of the lower sixth belt; and this belt also advances eastward in tabular form hereabouts. The northern part of the segment, Fig. 26, forms a hilly ridge only three k. wide between the incised frontal valley of the Seille on the east and the broader valley of the Moselle; the ridge is breached 20 k. from its northern end, where the Seille flows nearest to its eastern base. The villages of Ste. Geneviève (326 m.) and Mousson (380 m.) occupy the hilltops on the opposite sides of the breach; below the latter is the town of Pont-à-Mousson, on both sides of the Moselle. The fortified city of Metz (68,598; in French, pronounced *Mess;* in German, *Metz*) borders the river north of the cut-off segment.

The main body of the fifth upland back of the cut-off segment, like the whole body of the upland farther north, is more or less indented by the ravines of a number of short east-flowing streams, and is cut throughout by two longer streams which lead the drainage of the broad depression on the west through winding northeastward gorges, deepening as the upland rises, to the Moselle. One of the longer streams, named the Rupt de Mad, will be further referred to below in connection with the fourth upland belt; the more northern stream, the Orne, has a length of 50 k. and joins the Moselle not far north of Metz; its narrow, deepening, and singularly serpentine

FIG. 26. THE FIFTH UPLAND BELT AND

LLEY OF THE MOSELLE, SOUTH OF METZ

gorge crosses the frontier in the long western slope of the upland belt. These oblique gorges are too narrow and sinuous to be followed by main roads, several of which cross over the upland; but each gorge is followed by a railway and a secondary road.

33. *The Frontier on the Fifth Upland Belt.* The irregular and apparently arbitrary course of the frontier with respect to topographic features is strikingly illustrated in this district. It follows the winding course of the Seille through the tabular eastern extension of the sixth upland belt, as already stated; then as that stream turns north along the narrow part of the third upland segment, the frontier passes obliquely over the segment, crosses the open valley of the Moselle north of Pont-à-Mousson, traverses the crest of the main body of the fifth upland by a most rambling course, and turns northward on its western slope, as shown in part on the detailed maps, pp. 74, 75, and 78, 79, and also in Fig. 28.

The fortified city of Metz, German since 1871, lies as above noted on the Moselle (170 m.) just north of its exit from the incised valley; here the upland (350 m.) is strong and intact; the map on pages 78 and 79 shows its cleared upper surface, with its scarp, irregularly frayed out and forest covered, from the summits of which an extensive eastward view over the Lorraine plateau may be gained; the map also shows the low bench fronting the scarp base and the open floor of the Moselle valley, where an ancient Roman road still in use, a highway, and a railway are laid between the bench and the river. The river, after receiving the Seille at the northern end of the detached segment of the upland, flows 30 k. northward near the base of the bold upland scarp, before bending to the northeast to flow through the sixth upland belt, as will be further described in the account of the Lorraine plateau in chapter XIII. Thionville (German, *Diedenhofen*) lies at the bend (148 m.); the fifth upland belt (400 m.) here presents a scarp with the unusual height of 250 m.

The frontier hereabouts, although truly irregular, is not arbitrarily located: its course over the fifth upland belt was determined after the Franco-Prussian war of 1870 by the advice of a German geologist, who pointed out that this upland is rich in valuable iron ore (minette), of especial importance in the Bessemer process of steel making, and that by shifting the frontier from the base of the upland scarp, where it was at first proposed, a few kilometers to the west a large share of the ore beds would be transferred to Germany. For this reason the frontier, after crossing the Moselle between Pont-à-Mousson and Metz and ascending the upland scarp, runs irregularly northward on the western slope of the upland; from the mines there located a great share of Germany's iron supply was drawn before 1914.

Many smelting furnaces have been erected in the Moselle valley, the coal used there being brought from the Sarre basin in the Lorraine plateau to the east, to be described in a later section. To France was left only a lower and more western part of the iron-ore beds, known as the Briey area, from a town situated on the western slope of the upland. The entire iron-ore field of this area also has been in possession of Germany since August, 1914, and has made it possible, even with war in progress, to double her previous enormous production of iron and steel.

34. *The Woëvre Lowland.* The lowland between the fifth and fourth upland belts, for 80 k. north of the elbow of the Moselle at Toul, expands to a notable breadth because the belts are here wide-spaced, and receives a special name, *le Woëvre.* The maximum width of the lowland is 15 k.; its gently rolling surface has an altitude of about 230 m.; many artificial ponds are held in its shallow valleys; extensive forests overspread its low hills: its central area is, as above noted, drained eastward through the fifth upland belt to the Moselle by the Rupt de Mad and the Orne. This lowland will be referred to again in connection with Verdun, in the fourth upland belt.

FIG. 27. THE FIFTH UPLAND BELT AND T

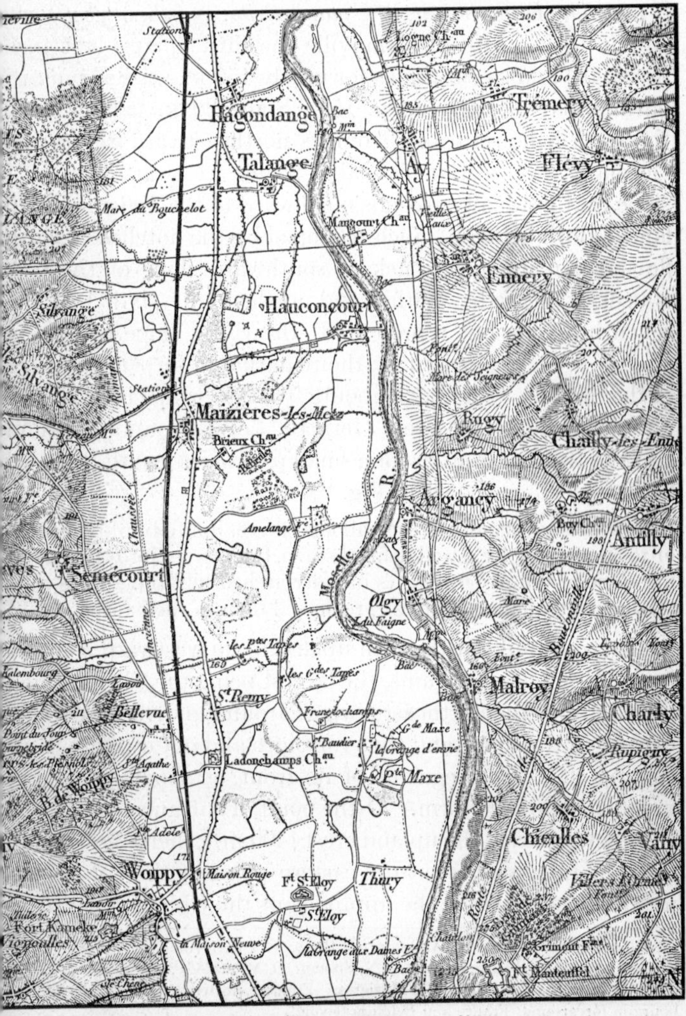

LLEY OF THE MOSELLE, NORTH OF METZ

35. *The Fifth and Sixth Upland Belts: Northernmost Parts.* About 10 k. north of Thionville the fifth upland belt turns westward and so continues for nearly 150 k., as shown on the outline maps, p. 69 and p. 153. Some 30 k. farther north the sixth belt, returning from its northeastward detour into Germany (see section 71), also trends westward along the slope that gradually ascends northward to the Ardennes highland. Both belts are much scalloped by the notches of southflowing streams: the back or southward slope of the belts is usually cleared. Within and north of the sixth belt (the seventh belt is not represented) the rising slope of the Ardennes is deeply incised by the remarkably serpentine valley of the Semois. The Luxembourg frontier, which is here interposed for 8 k. between German Lorraine and Belgium, and the Belgian frontier for a small part of its length lie on the fifth upland, as shown in Fig. 28.

The sixth upland belt hereabouts is not limited on its exterior or northern side by a well defined scarp, but merges into the ascending slope of the Ardennes; its broadly arched hills differ from those farther north more in their calcareous soil than in their form.

The continuity of the fifth upland in its westward course is broken by several streams, the largest being the Chiers which, coming from the northeast, swings around an irregular course convex to the south and thus runs into and out of the upland, cutting off a 40-k. segment of its front, as shown in Fig. 28 in much simplified form. At the point of entrance, the stream is 140 m. beneath the upland crest (400 m.); here the fortified town of Longwy lies on the eastern point of the cut-off segment and commands the approach of the Chiers valley from

Abbreviations in Fig. 28 : A, Arlon; B, Briey; C, Charleville; D, Dun-sur-Meuse; E, Etain; L, Luxembourg; M, Metz; N, Stenay; O, Montmédy; S, Sedan; T, Thionville; V, Verdun; Y, Longwy; Z, Mézières. Frontier, dotted: Germany and Luxembourg in foreground, France and Belgium beyond.

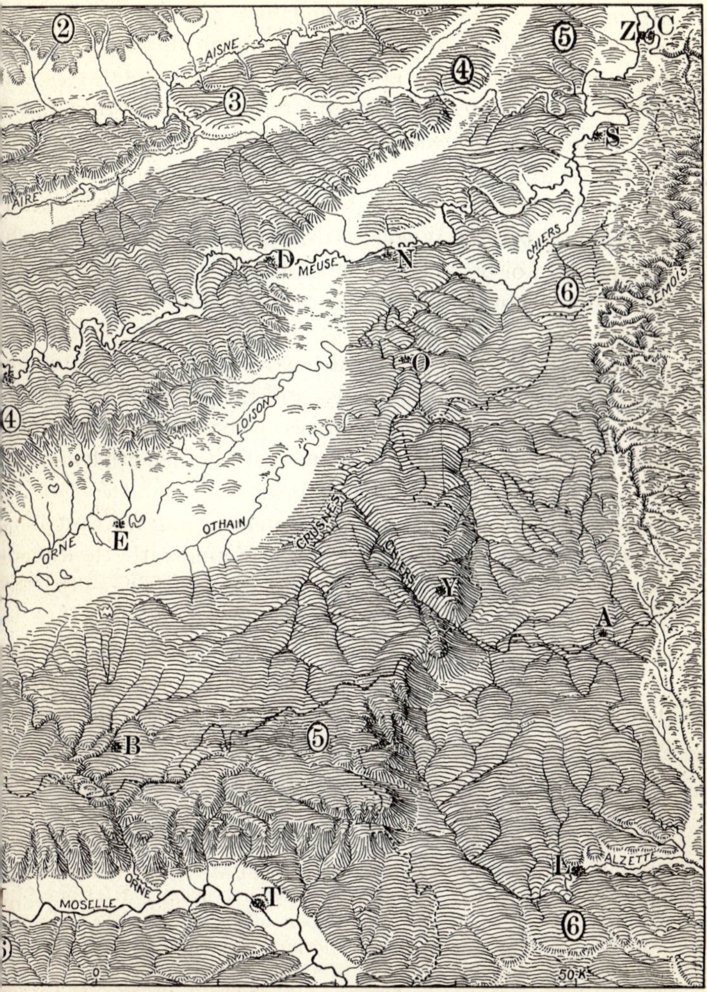

28. Upland Belts South of the Ardennes, looking West

Luxembourg. The valley of the Chiers through the upland wanders irregularly; Montmédy lies on one of the valley-side spurs in the upland not far from the river exit.

The back slope of the fifth upland belt, at its turn from a northward to a westward trend, north of the Briey iron-ore district, is drained by the Crusnes, which joins the Chiers at the southernmost point of its segment-cutting curve. The northern part of the Woëvre is drained northwestward by the Othain and Loison to the Chiers; like the Chiers, all three of these branch streams cut narrow, winding valleys in the back slope of the upland.

Farther west, the upland, much narrower than near Metz, is obliquely trenched by the Meuse (165 m.), flowing northwest; there Stenay and Mouzon lie between the hills of the upland (350 m. on east, 330 m. on southwest): it thus appears that the Meuse, which flows obliquely inward through the fifth upland belt between Langres and Neufchâteau, flows obliquely outward through it south of the Ardennes; the points of entrance and exit are about 180 k. apart. Evidently the northernmost part of this upland belt, reduced in width to about 10 k., has little continuity: nevertheless the form and the relative positions of its parts are best appreciated when they are recognized as belonging together although cut apart by traversing rivers.

Mouzon is notable as marking the crossing place of an ancient Roman road, which holds an almost direct course over hill and dale, still followed for long stretches by secondary modern roads, between Rheims, 85 k. distant to the southwest, and Trèves (German, *Trier*) on the Moselle, 110 k. distant to the east-northeast. After crossing the high crest of the fourth upland belt (336 m.), the road descends and traverses the lowland (here no longer called the Woëvre) between the fourth and fifth belts to Mouzon; it then passes over the fifth upland to the next lowland, where Carignan lies at the crossing of the Chiers; then along the sixth upland belt and away. Mouzon and Carignan are therefore good examples of ancient river-crossing towns, the general location of which is dependent upon the intersection of an almost direct long-continued road with the rivers on its course, though the precise sites of crossing were probably influenced by suitable points for fording, at which bridges were later built as the towns grew in size.

After its exit from the fifth upland belt, the Chiers flows westward on a broad flood plain through the lowland between the dissected scarp of the fifth belt on the south and the long back slope of the sixth belt on the north. Similarly the Meuse, after issuing from the fifth belt and receiving the Chiers, wanders on the same wide flood plain as it flows westward through the lowland. Sedan of fateful memory lies six k. below the junction of the two rivers. The heights of the fifth belt (346 m.), here much narrower than farther east, rise rapidly on the south; the cleared back slopes of the scalloped sixth belt (310 m.) ascend slowly on the north. Next west of Sedan, the Meuse makes a strong northward loop into the back slope of the sixth upland belt; farther west it makes a double loop, where the fortress of Mézières and the city of Charleville lie near each other on the lowland (150 m.) with the river between them; then the river turns north, cuts through the sixth upland belt (280 m.), and enters a deep and winding gorge which it follows through the Ardennes. The two upland belts weaken and disappear a little farther west.

CHAPTER VII

THE REGION OF THE MEUSE

36. *The Fourth Upland Belt: Southern Part.* The fourth upland belt has a greater length and in its middle a greater breadth than any other; it is recognizable, though imperfectly developed, along the northern border of the Central Highlands west of the broad valleys of the Loire and the Allier; while to the east of those valleys it is a well defined lineament, although cut apart by many streams, through all the 350 k. of its rounded northward course, Fig. 24, which leads it nearly to the Ardennes. Its limestones are less pure and their forms are less bold than those of the fifth belt; its cross profile as a whole is gracefully but unsymmetrically arched; its east-facing scarp is delicately scalloped and its gray frontal slopes are generally cleared; its broad upland and long descending slope are usually carved into rounded hills, many of which are too dry for repaying cultivation and are therefore largely left to tree growth.

Railways follow nearly all the many cross-valleys, of which the northernmost is that of the Marne, that traverse the southern part of the fourth upland belt; hence these railways, river-like, converge and unite on their way toward the metropolis at the center of the upland arcs; but here a circumferential railway also is seen, following the inter-upland depression between the fifth and fourth belts through the southern half of its long curve, because of the importance given to it by its breadth, for the crests of the fifth and fourth belts are 40 k. apart in the south. The depression between them is moderately trenched across by all the rivers that traverse it.

As a result of the widening and eastward advance of the fourth upland, the two belts converge northeastward and the depression between them narrows until their crests are separated by only 14 k. where the five close-set eastern branches of the Meuse, mentioned in section 30 as trenching the fifth upland, unite with the main stream, which in its further northward course obliquely enters the fourth upland. Here, on the dissected back slope of the fifth belt, bordering the narrowed depression, Neufchâteau (310 m.) is picturesquely situated, 20 m. above the level of the incised streams and 130 m. below the crest of the fourth upland which rises abruptly a short distance to the northwest. The features of this district are therefore typified by the trenched depression at the right end of the upper diagram, p. 47.

Neufchâteau is a center for a number of radiating railways and highways, which run in pairs along the circumferential depression between the fifth and fourth upland belts southwestward to Chaumont-en-Bassigny on the Marne, and northeastward to Toul and Nancy; also up the Meuse, thus passing through the fifth upland belt and joining other roads by the Marne beneath Langres; and down the Meuse to the main line of the Eastern railway west of Toul; also southeastward up a branch of the Meuse on the way to Mirecourt in the sixth upland belt and beyond; and a railway crosses over the fourth belt west and northwest to the Ornain.

37. *The Fourth Upland Belt: Northern Part.* Next north of Neufchâteau the fourth upland belt attains its greatest breadth of 50 k. Here the valley of the Meuse obliquely enters the main body of the upland and 110 k. farther north flows obliquely out again, thus cutting off a long segment, which is known where its breadth is greatest as the *Côtes de Meuse.* This part of the Meuse valley may therefore be likened to the chord of an arc, but it has nothing of the single-minded directness of a geometrical line: its general trend to the north-northwest is bent eastward to the mouth of the Val de l'Ane where the Moselle formerly flowed west, before it was diverted at the elbow of capture by Toul. All along this

course the valley is gracefully serpentine, while the river has an even more elaborately sinuous course in the smooth meadows of the serpentine valley floor. Through all this distance the main body of the upland, west of the river, is not cut through by any stream. Indeed, from the oblique trench of the Marne, the main body of the upland is not transversely trenched for 170 k.; but near its farther end, where the crest declining northwestward, has altitudes of about 300 m., it is traversed by a small stream, the Bar, as will be further stated below.

The main body of the upland, has, nevertheless, a varied form. Its eastern slope is deeply indented by several short east-flowing branches of the Meuse, south and north of the former junction of the upper Moselle. The broadest part of its western slope, opposite the first half of the cut-off segment, is overlapped by a subordinate upland, with a greatly dissected front nearly 100 m. in height shown in part on the detailed map, page 94; and the back slope of this upland, elaborately carved into rounded hills by many small streams, is obliquely incised for distances of 30 k. by the Ornain, a branch of the Marne, and by the Aire, a branch of the Aisne-Oise both flowing northwest. Transverse transportation lines must therefore either climb over or tunnel through the northern part of the fourth upland belt.

The main line of the Eastern railway — Chemin de Fer de l'Est — from Paris follows up the Marne and the Ornain to the western slope of the fourth upland belt, where Bar-le-Duc (181 m.) lies on the last named river; the railway continues for a short distance southeastward and then turns northeastward and crosses over the upland at an altitude of 325 m. a broad view is gained from the treeless crest, before descent is made to the open valley of the Meuse (240 m.); this is followed southeastward to the Val de l'Ane, through which the railway turns east to Toul on the way to Nancy; other lines branch up the Meuse to Neufchâteau (section 36) and down the Meuse to Verdun and beyond.

A secondary line and the Marne-Rhine canal continue up the Ornain farther than the main line of the Eastern railway, following a part of the river where it flows in a minutely sinuous channel along an incised meandering valley; the railway continues southeastward over the upland to Neufchâteau: the canal turns eastward, tunnels 4 k. through the upland to a small valley of the eastern slope which it follows down to the Meuse; beyond that river the canal turns, like the main railway line, through the Val de l'Ane to Toul; thence down the new course of Moselle and up the Meurthe to Nancy on its farther way, which has been described in section 25.

38. *The Underfit Meuse.* The uncertain, hesitating course of the Meuse through the flood-plain meadows of its meandering valley deserves further mention. It runs anywhere but around its valley curves; it is thus unlike the lower Seine, which swings vigorously around and fits closely into the large curves of its meandering valley through the chalk uplands of Normandy; and unlike the lower Moselle, which in similarly well-ordered fashion fits the meanders of its narrow valley, deeply incised in the Slate-mountain highlands on the way to the Rhine. The elaborately sinuous Meuse may therefore be described as "underfit," in the sense of being incompetent to follow its valley curves; it thus illustrates a curious habit, widely prevalent among the smaller rivers of northern France, such as the Vézouse, the upper Meurthe, the Chiers, and the Ornain, above mentioned, and the Aisne, the Oise, and the Somme, to be described below.

It has been suggested that the underfit habit of these rivers is due to loss of river volume by slow underflow in the deep alluvium — gravels, sands, and silts — of the valley floor; thus it is implied that at an earlier stage of development, before the alluvium of the flood plain was deposited, the visible river carried the total drainage of its basin in its channel, and was for a time large and vigorous enough to cut out the curves of a freely meandering valley, into which it then necessarily fitted because they were its own product; but later, as alluvium began to accumulate on the rock-bed of the valley, the visible volume, flowing in the surface channel, was decreased by the amount of creeping underflow in the alluvium; the deeper the alluvium became the more the visible river flow was thus diminished, until at the present stage of its development it is reduced to so small a current that it is altogether incompetent to swing around the valley curves, which it had itself previously carved, as is clearly shown on the detailed maps, pp. 90 and 95.

The Meuse may therefore be taken as the type of an underfit river; its course is so much more serpentine than that of the valley that it sometimes, as at points near St. Mihiel and Verdun on the detailed maps above noted, flows almost backwards towards its source, as if, with loss of volume, it had also lost the sense of direction and knew not where to turn! The practical application of this explanation is that serpentine valley floors, on which "underfit" rivers wander irregularly, may be expected to have a deep alluvial deposit burying their rock bottom.

39. *St. Mihiel and Verdun.* The upland segment which, cut off from the main body of the fourth upland belt by the Meuse valley, forms the Côtes de Meuse, is breached at the first quarter of its 100 k. length by the winding Val de l'Ane, already described as connecting the elbow of the Moselle with the valley of the Meuse, and as marking the former path of one river to its confluence with the other: this dry valley today serves as a gap for highway, canal, and railway as above noted; the first crosses over one of its valley-side spurs under which the other two pass in tunnels.

A little farther north the northeast-flowing Rupt de Mad, one of the branches of the Moselle, briefly mentioned in section 32 as trenching the fifth upland belt, has pushed its head so far westward across the Woëvre lowland by retrogressive erosion as to excavate a bight or concave reëntrant in the scarp of the Côtes de Meuse, and thus reduce them to their least width of three k., as shown in the lower part of the map, pp. 90, 91. Just here, moreover, two small west-flowing branches of the Meuse have cut their valleys almost through the narrowed Côtes.

Hence it was here that the German forces early advanced westward from the fortress of Metz across the fifth upland belt and the Woëvre lowland, and reached the Meuse in a salient between the fortresses of Toul on the southeast and of Verdun on the north; the apex of the salient is a little north of the narrowest part of the upland segment at mid-length of the Meuse valley-chord, and is marked by the town of St. Mihiel: there on a west-reaching spur that enters the second turn of an S-like pair of valley curves, a French stronghold had been built upon the site of an ancient Roman camp, but unfortunately on the east side of the river; the Germans captured the stronghold and cut the railroad in the valley beneath. Thus defended on three sides by the natural valley-moat, they have held possession against all French attacks. Their object was plainly enough not merely the occupation of this point, of relatively small value in itself, but the isolation of Verdun, farther north in the Meuse valley, of which more is told below.

North of the Rupt de Mad bight, the segment of the Côtes de Meuse (380 m.), partly forested, partly cleared, is high and continuous in the sense of not being cut through by any transverse valley, but it is carved into many rounded hills separated by deep ravines. The east-facing scarp of the segment is not so sharp crested as that of the fifth upland belt; nevertheless, it is a striking topographic feature. The view eastward from its promontories includes the whole breadth of the Woëvre with patches of shining ponds, and the long western slope of the fifth upland. As is often the case elsewhere, so here the level of emerging springs on the lower slope of the scarp

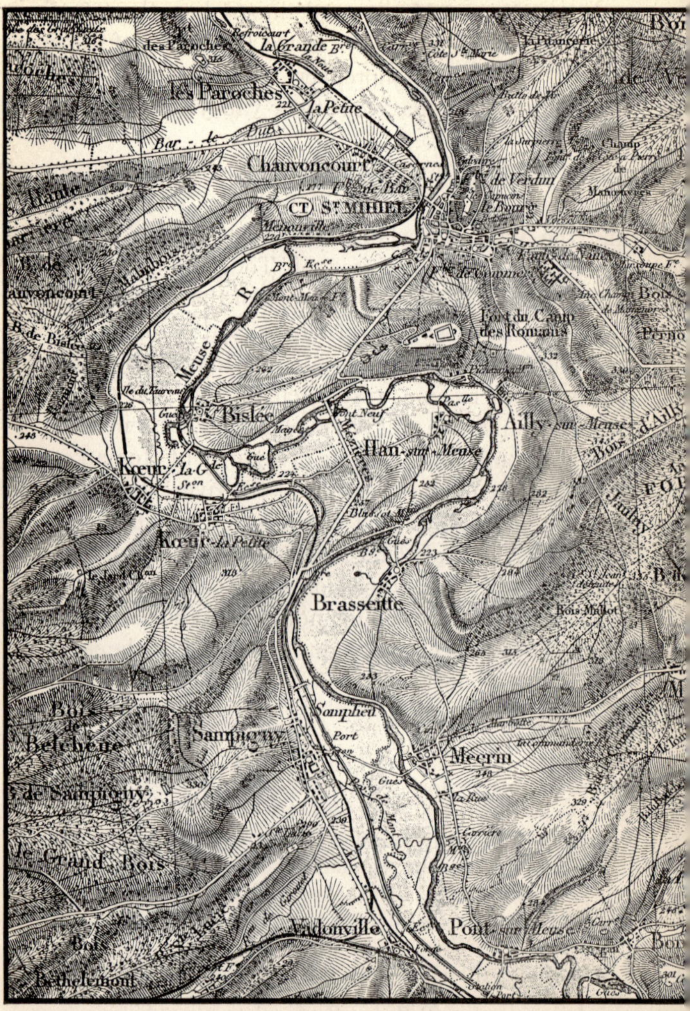

FIG. 29. THE MEANDERING VALLEY OF THE M

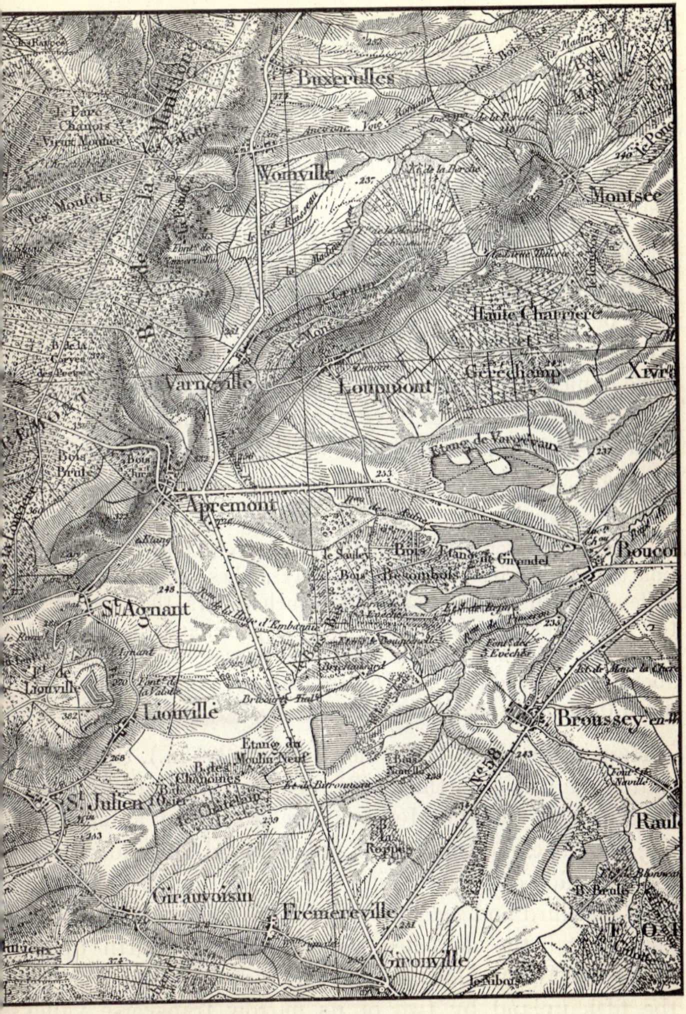

THE FOURTH UPLAND BELT, NEAR ST. MIHIEL

determines the site of a series of ancient villages which have been there from time immemorial; the villagers are in large measure descended from a long line of local forefathers; their number changes slowly. Many of the villages have *sous-les-Côtes* appended to their name. In proud contrast to the humbler position of these lowly villages, Hattonchâtel (-castle) occupies a commanding promontory of the segment crest, and recalls an era of barons lording it over peasants.

The main western body of the upland belt hereabouts, west of the Meuse, is still a broad and hilly upland (320 m.), in spite of the loss of its higher frontal segment. The relief is so strong on both sides of the river that the upland villages often have names ending in *-mont*, as Haumont, Louvemont; some of those in the ravines have names ending in *-court*, as if to indicate their narrow enclosure by the hills, as Chattancourt, Avocourt, Landrecourt.

The meandering habit of the Meuse valley through the hills results in systematic differences of length in the several lines of transportation that follow it; the main highway makes short cuts over the valley-side spurs, and is, as usual in such valleys, the shortest line; the railway along one side of the flood plain, and the canal often along the other side, follow the valley curves; the river, twisting about on the flood plain, and occasionally even turning backward, is much the longest line of the four.

Here in the chord-valley lies Verdun (204 m.), the fortified center of a ring of fortified hills. The railway that follows the winding valley of the Meuse is here crossed by an east–west line that rises and falls over uplands and lowlands; to the west it leads over the main body of the fourth upland and over the third and second uplands to Rheims; to the east, it passes over the detached segment of the fourth upland, across the Woëvre lowland, and then dividing, runs through the fifth upland by two of its narrow transverse valleys.

Unfortunately all these railway lines have been controlled by the Germans since shortly after the beginning of the war.

Although thus cut off from supplies by rail, Verdun has successfully resisted all assaults. Farther down stream, the valley and the hills on both sides of it, as well as most of the Woëvre lowland on the east, were early occupied by the Germans; but it has been impossible for them to make successful attack upon Verdun from the lowland, by reason of the natural defense offered by the strong scarp of the cuesta segment; the effort of the Germans to ascend a ravine in the scarp face, known by the name of the village, Vaud, at its mouth, cost them the sacrifice of thousands of lives. It is therefore from the north, where hills about as high as those crowned by the forts around Verdun are occupied by the enemy, that the strongest attacks have been made; but these attacks have been impeded by the deep side-valleys which divide the hills.

Some of the hills have become notorious in the course of repeated attacks and repulses: Côte de Froide Terre (345 m.) is 4 k. north of Verdun; Douaumont (388 m.), a violently contested point, rises 9 k. to the northeast; le Mort Homme (295 m.) is 12 k. northwest of the city; near by on the west an advancing spur of the subordinate overlapping upland is known from its height as "Hill 304." The digging of trenches and the blasting of "craters" by innumerable shells has inflicted a long-lasting injury on the fields of the uplands, not only by making them uneven, but even more by mixing the surface soil with a great volume of unweathered rock fragments. The surface can be graded smooth again in a few years of peace, but it will require scores of years to restore its lost soils.

FIG. 30. THE MEUSE AT VE

he Fourth Upland Belt

It is evidently because of the separation of Verdun by the main body of the fourth cuesta from the more open country farther west, that it has played so individual a part in the war. It withstood the first assault of the Germans in August, 1914; since their retreat from the Marne in September of that year, the attacks upon Verdun, especially the long-continued assaults of 1916, have been essentially independent of the campaign farther west. The defense of the fortress has been made doubly difficult since the Germans took St. Mihiel, as noted above, and thus prevented the bringing of supplies by rail along the valley floor from the south. To overcome this deprivation, thousands of motor trucks, running on schedule time, have been used to bring munitions over the upland from Bar-le-Duc on the Ornain, by a hilly road that crosses the incised valley of the upper Aire on the way. The dislodgment of the Germans from St. Mihiel and the reëstablishment of railway communication along the Meuse valley would therefore greatly strengthen Verdun.

It is worth noting that by a treaty made at Verdun in the year 843 the vast empire of Charlemagne was divided among his successors; thus for the first time did the region which we know as Germany have a ruler of its own; the beginning of German national life may be placed at this date. It is not improbable that the frantic efforts made by the Germans to capture Verdun have been prompted by pride awakened by this historical reminiscence: but the French have said, " They shall not pass."

40. *The Fourth Upland Belt: Northernmost Part.* About 30 k. beyond Verdun the underfit Meuse leaves its winding valley in the fourth upland belt and flows out upon the northwestern extension of the frontal lowland, leaving the upland intact on the southwest. The lowland is here less than 10 k. in width and no longer bears the name of Woëvre. Dun-sur-Meuse lies on the river at the exit from the upland, and Stenay is beyond the lowland some 12 k. farther on.

The upland trends northwest, and rapidly losing breadth is cut through obliquely by the Bar, 30 k. northwest of the exit of the Meuse (see diagram, p. 81); it continues about as much farther (see map, p. 153), narrowing as it is crowded between the converging fifth and third uplands on either side, and then disappears under the overlapping chalk upland, as will be described in section 46. The ancient village of Stonne stands on the upland summit next east of the Bar valley, where the Roman road from Rheims to Trèves holds its undeviating way over the upland belt.

The Bar is a remarkable example of an underfit stream; its minutely sinuous course is about twice as long as the larger curves of its serpentine valley. Its minute sinuosity is doubtless due to diminution of volume, for which two reasons may be given: the Aire once continued its flow northward to the Bar before it was captured by a branch of the Aisne, as is further stated on p. 100; then the beheaded Bar, no longer able to transport all the detritus brought from the adjoining uplands by the side streams, laid down some of it on the valley floor, which was thereby built up or aggraded by a considerable thickness of alluvial deposits; and a further loss of surface volume was caused by underflow therein, as explained in section 38.

CHAPTER VIII

ARGONNE AND CHAMPAGNE

41. *The Third Upland Belt: its Southern Lowland Substitute.* The third upland belt is peculiar in being well developed only in the northern part of its arc; farther south the determining strata of this upland are thin or absent, and its place is taken by low rambling hills (180 m.) bearing large forests, or by a lowland (140 m.), 20 k. in width, between the long back slope of the fourth upland belt and the low front of the second, as shown in Fig. 32. The main branches of the Seine system cross the lowland in open flood plains. Indeed, in the space between the Marne and the northward bow of its chief branch, the Ornain, several smaller streams, gradually converging westward, aid in transforming almost all that part of the lowland, over a north–south stretch of 20 k., into contiguous plains, on which the streams wander with little restraint.

The larger towns and cities of this district avoid the lowland and occupy the main-stream valleys to the southeast or northwest. Thus Bar-sur-Seine, Bar-sur-Aube, St. Dizier on the Marne, and Bar-le-Duc on the Ornain are at the edge of or within the back slope of the fourth upland belt; while Troyes on the Seine and Vitry-le-François on the Marne, regarding both of which more is said below, occupy reëntrants in the low front of the second upland belt. Numerous roads and railways traverse the lowland in various directions; the most important railway is the main Eastern line, which crosses from Vitry-le-François along the Ornain to Bar-le-Duc.

42. *The Forest of Argonne.* To the north of the Ornain, a number of large ponds are held among the lowland hills, as is

also the case farther south between the Marne and the Aube. The hills then increase in height and soon begin to assume a cuesta-like form; it is not, however, until 20 k. beyond the Ornain that the third upland belt, maintained by beds of argillaceous sandstone, gains a well developed relief with a width of 15 k. between the Aire on the east and the Aisne on the west, both of which flow north-northwest in exceptionally well directed longitudinal courses. The upland begins rather abruptly, thus presenting a descent to the south; thereafter it has normal form, with a fairly strong slope falling off to the east, and a longer slope descending to the west; it thus continues for 75 k. to the north-northwest, with stony, infertile

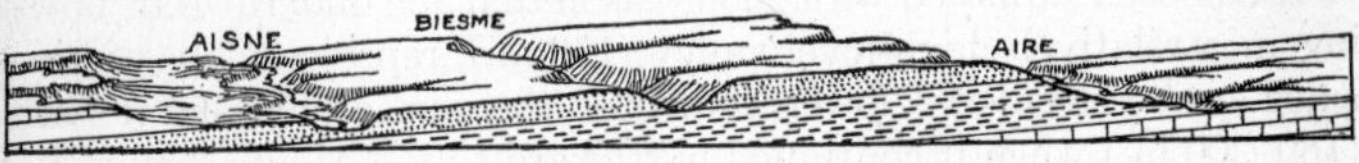

FIG. 31. SECTION ACROSS THE FOREST OF ARGONNE

soil; much of its length is covered by the great forest of Argonne. The district of Argonne extends eastward and includes part of the back slope of the fourth upland.

A characteristic profile across the southern part of this forested upland may be begun in the lowland (240 m.) on the east. Beneath it, the Aire has incised a meandering valley of small depth, near which Clermont-en-Argonne and in which Varennes-en-Argonne are the chief towns; the upland crest rises to a height of 308 m., whence a long westward descent (200–170 m.) leads to the next lowland (150 m.), before reaching which, however, one must cross the longitudinal valley of the Biesme which splits the upland for some 20 k., and the meandering valley of the Aisne (140 m.); the latter is an underfit river in a meandering valley near the base of the slope. Ste. Menehould is the chief town in this part of the Aisne valley. A transverse railway, coming from Rheims, passes Ste. Menehould, crosses the split upland to Clermont-en-Argonne, and then traverses the main mass of the fifth upland belt to Verdun.

At mid-length of the forested belt, the upland is cut square across by the valley of the Aire, which there leaves the longitudinal lowland east of the cuesta and turns to a transverse course; Grand Pré lies close to the elbow where the turn is made. The following part of the lowland is drained northward by the Bar, an extremely underfit stream, the lower course of which follows, as above noted, a meandering valley of large pattern through the fourth and fifth upland belts.

It has been supposed with good reason that the underfit Bar, flowing in a relatively high-level valley (160 m.), represents the diminished or " beheaded " lower course of the former high-level Aire (180–200 m.) when it continued northward to the Meuse, before its diversion to the Aisne in the lower western lowland (120 m.); hence the bend of the Aire at Grand Pré may be regarded as an elbow of capture, like that of the Moselle at Toul; and the incision of the Aire valley (140 m.) may be explained as a consequence of the transfer of its allegiance from the higher levels of the Meuse system to the lower levels of the Aisne-Oise-Seine system; but the valley of the little Aire at the Grand-Pré elbow of capture is much wider than that of the larger Moselle at Toul; hence the capture of the Aire by a branch of the Aisne should be regarded as more ancient than that of the upper Moselle by a branch of the aboriginal Meurthe.

It is worth noting that the Aisne below the point where the Aire has been added to its volume, and where it therefore might expectably show the vigorous habit of an " overfit " river, more competent than ever to flow vigorously around its valley curves, is nevertheless strikingly underfit; hence loss of volume by percolation in flood-plain alluvium, as above suggested, is a plausible explanation of its now enfeebled behavior.

Vouzier (100 m.) lies in the Aisne valley below the confluence of the Aire. Not far beyond that town and 23 k. below the confluence of the Aire, the Aisne turns westward; its further course will be described in a later section. Opposite this turn, the third upland belt, here trending northwest, is almost cut through by a small branch stream, the cross valley of which is ascended by a canal which connects the Seine-Oise-Aisne river system with that of the Meuse; the canal makes a short-cut path through a spur-stem of the Bar valley near its junction with the Meuse. A modern lane follows the ancient Roman road, above described as connecting Rheims and Trèves, over the hills of the third upland belt (240 m.) next south of the cross valley just mentioned. Beyond the cross valley, the lowland in front of the upland narrows and disappears, presumably because its determining weak strata give out; the third upland thereupon almost merges with the fourth, as already stated.

43. *The Second Upland Belt, Southern Part: The Forest of Othe.* The second upland belt, formed of chalk strata, is a low and much scalloped bench (190 m.) rising gently with gracefully curved profile from the northwestern side of the broad lowland (140 m.) that south of the Marne replaces the third upland belt; but southwest of the Seine the bench is dwarfed by a higher upland of cuesta form (280 m.), covered by the forest of Othe, which rises with a strong frontal slope, deeply scored by steep ravines, a few kilometers further northwest. This high upland is maintained by a body of sandstone strata, which singularly enough do not extend northeast of the Seine, and which lose topographic value southwestward of the Yonne; but between these limiting rivers the upland that they form is a dominating feature.

44. *The Second Upland Belt and the Champagne.* North of the Forest of Othe the second or chalk upland, as it may be called, although of well defined form, is so low that it does not strongly separate the broad lowlands on its two sides. A large part of this extensive region of small relief, including most of the broad lowland where the third upland is wanting,

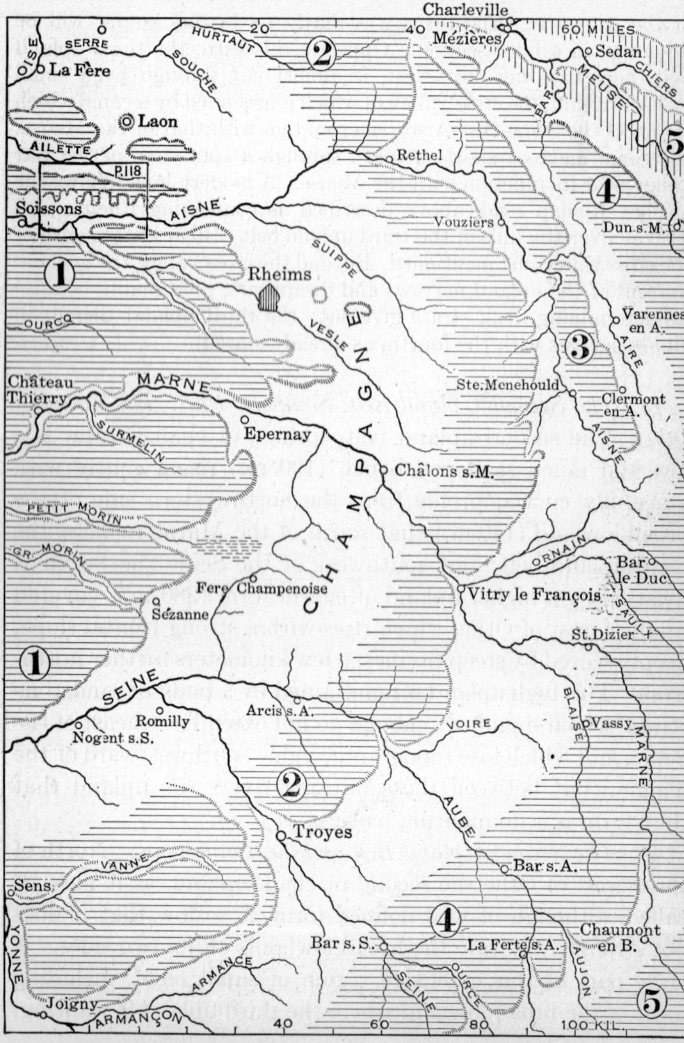

FIG. 32. THE FIRST, SECOND, AND THIRD UPLAND BELTS

the low second upland, which bears many small patches of woodland, as far as the Aisne, and a long stretch of the following lowland of sandy and clayey strata which fronts the strong scarp of the first upland belt, is known as *la Champagne*, or "the open country."

Of this area, the broad eastern lowland as far north as the Forest of Argonne (third upland belt), and the narrower northward extension of this lowland between the Forest of Argonne and the chalk upland, are characterized by many streams, and by moist, deep-soiled fields and numerous villages; these parts are therefore united under the name *la Champagne humide*, or the moist Champagne. The western part, including the low chalk upland and the following lowland of few streams and dry soils, where villages and cultivated fields are limited to the wide-spaced valleys, has received the unflattering name of *la Champagne pouilleuse*, which may be called the dry Champagne: it is this drier western area that is now to be especially considered; and to it, rather than to the moister eastern area, the name Champagne usually applies.

The southwesternmost extension of the dry Champagne lies in the lowland between the long back slope of the Forest of Othe and the strong scarp of the first upland belt, where the Seine, having received the Aube, flows 60 k. west-southwest and receives the Yonne before resuming its northwest course and passing through the first upland to Paris. From this beginning the belt of low or moderate relief, widening by the addition of the low chalk upland belt north of the Seine, sweeps around an arc of more than 90° and over a distance of 250 k., from the Yonne on a southeastern radius of the Paris basin to beyond the Aisne, which limits the Champagne district, and as far as the Oise on a north-northeastern radius; the undulating lowland surface continues west of the Oise into northernmost France, as will be described in section 52. This belt of open country, including the low chalk upland on the east along with the lower lowland of sands and clays to which the chalk upland gently declines on the west, is much more extensive than any other physiographic area of northeastern France. For convenience of description it may be divided into sectors by the transverse rivers; the sector beyond the Oise will be treated on later pages.

45. *The Dry Champagne from the Seine to the Aisne.* The Seine flows through the second or chalk upland belt next north of the Forest of Othe in a broad and squarely transverse valley, on the floor of which lies Troyes (55,486), the largest city between Paris and Dijon. It is from the extensive dealings in the ancient fairs of this commercial city that English-speaking people have acquired the so-called Troy weight, in which 12 ounces make a pound. Beyond the open breach of the Seine, the chalk upland continues its generally treeless surface for 160 k.; yet low as it is only three rivers, the Aube, the Marne, and the Aisne, traverse it in all this distance. At the entrance to the obliquely transverse valley of the Marne lies Vitry-le-François, one of the most southern points reached by the German army in its first advance. The lowland of the moist Champagne (140–150 m.) to the east of the chalk upland, unusually broad through the long 100 k. stretch where the third upland belt is wanting, is much narrowed, as above noted, after the rise of that upland in the Forest of Argonne farther north; the narrowed part of the lowland is, singularly enough, not followed by a longitudinal stream, for it is in the back slope of the third upland adjoining on the east that the underfit Aisne pursues the incised meandering valley described in section 42.

The frontal scarp of the chalk upland is elaborately carved by the close-set ravines of many short, east-flowing streams, the branches of which head two or three k. back in the upland; its generally treeless spurs are unlike those of any other upland belt in the series from the Vosges to Paris. Villages lie in the ravine mouths or beneath the spur ends, where a water supply is obtainable; but they are more abundant on the better watered and well cultivated lowland farther east, in its

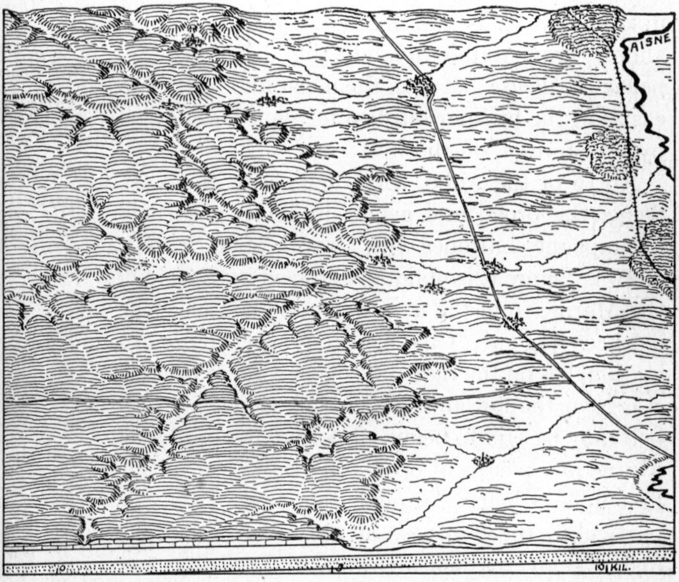

FIG. 33. THE RAGGED SCARP OF THE LOW CHALK UPLAND

narrowed northern extension parallel to the Aisne, as well as farther south where it is broad. The chalk strata that constitute the second upland belt (200 m.) are so pervious to water that the thin soil over the flat upland crest is unusually dry; hence villages here are few and far between: even the valley heads of the back slope are waterless. Not until the slope declines some 40 or 50 m., on the western side of the faintly convex upland crest, do the shallow valleys bear streams; and not until after the streams begin are villages found also. The roads of the upland are surfaced with flints. Thus to the easily recognized flint-bearing chalk of its maintaining strata and to its well individualized though low relief, the second upland belt adds the characteristic of a thin-soiled, dry, almost uninhabited upland, the driest part of *la Champagne pouilleuse*, between the moist and fertile lowland on the east and the broader but less fertile lowland, partly occupied by pine forests, on the west.

North of the Aube, which curves to the west to join the Seine, the broad and gently undulating surface of the dry Champagne continues with a width of 40 or 50 k. for some 60 or 70 k. before it is obliquely crossed by the shallow valley of the Marne, from the marshy flood plain of which much peat has been dug out. Châlons-sur-Marne, an important military station, lies on the right side of the valley, where the eastward rise from the lowland to the back slope of the second upland belt may be said to begin. Epernay lies on the left side of the Marne in a reëntrant opened by that river in the front of the first upland belt. Through this part of the Champagne in particular, the faintness of the relief is shown by the long distances over which highways and railways run in straight courses.

Then follows a northward stretch of some 60 k. where, between the Marne and the Aisne, several small streams rise in shallow valleys among the woodland patches on the gentle back slope of the chalk upland and cross the lowland northwestward to the Aisne. Here the battle front has lain for three years, trending from the southern part of the Forest of Argonne west-northwest to the Aisne where it enters the first upland belt. The district has been the scene of severe fighting, with moderate northward gains for the French. The shallow valleys of the Vesle and the Suippe in the rolling surface are of less strategic value than several residual hills in the center of the area between these streams, on which the Germans entrenched their forces after retreating from the Marne. The hills are composed of strata overlying the chalk; the highest of them (267 m.), eight k. east of Rheims, may be named from the village of Berru on its eastern slope. Twelve k. farther on other hills (257 m.) rise west of the village of Moronvilliers. Their possession has been desperately contested. Tunnels driven through the hills gave the Germans protection in passing from their camps on the sheltered northern slope to the trenches on the exposed southern slope. When the hills were captured in the spring of 1917 by the French, after heavy artillery firing by which the tunnel mouths were broken down and obstructed, hundreds of German soldiers were found suffocated in the tunnels. The location of the hills is shown in the diagram on pages 112, 113.

In the western part of this division of the Champagne north of the Marne and beyond a strong eastward salient of the first upland belt, lies the famous cathedral city of Rheims (French, *Reims;* the final *s* is pronounced) on the the Vesle, a small branch of the Aisne. Since the retreat from the Marne, the German line has been held not far northeast of the city, which has thus been exposed to intermittent bombardment for over three years. The ancient origin of this city is attested by the number of Roman roads that radiate from it: one of them, already mentioned as leading to Trèves, crosses the plain northeastward in a remarkably direct course.

46. *The Lowlands from the Aisne to the Oise.* The Aisne, turning west from its longitudinal course in the back slope of the third upland belt, crosses the next narrow lowland, the low chalk upland — Rethel lies here in the transverse valley — and the lowland plain beyond, and then enters the plateau which constitutes the northern extension of the first upland, as will be explained below. In its course west of the chalk-upland crest, the river flows in a very sinuous channel through a wide and often marshy flood plain. To the north of

the Aisne (see maps, pp. 102 and 153) the remaining portion of the undulating southwestward slope from the second or chalk upland to the adjoining lowland forms the northernmost sector of the broad belt of small relief that began at the Seine. The belt here extends beyond the limits of the Champagne district and reaches the Oise. The drainage of the area is accomplished by the west-flowing Souche, Hurtaut, Brune, and Villepion, all small streams that unite in the Serre and thus reach the Oise, which in this upper part of its course has, like the Aisne above mentioned, a winding channel in a marshy flood plain.

The southwestern and lower part of this northernmost sector, near the group of plateau segments that here represent the first upland belt, the smallest and northernmost member of which is crowned by Laon, is so low and flat that it is drained by artificial canals. The northeastern part of the sector, rising gradually with the northward ascent of the chalk formation, has deeper valleys and therefore a stronger relief than the corresponding area in the Champagne; its many broadly rounded upland hills rise among a labyrinth of valleys, with neither height nor depth enough to make movement difficult in time of peace, yet with such variety of form as to give in war much advantage to well chosen lines of defence and to impose corresponding disadvantage on the lines of attack.

The northward extension of the undulating uplands is peculiar. For a short distance north of the Aisne, they fall off eastward, as heretofore, in a scarp of moderate height which descends to the narrow lowland already described, and beyond the lowland a moderate ascent leads up the back slope of the weakening third upland; but farther north the scarp disappears and the chalk upland successively overlaps the lowered ends of the third, the fourth, and the fifth upland belts (see map, p. 153), and farther west it wraps around the westward slope of the Ardennes into Belgium. Thus the strong rampart-like scarps of the several upland belts, which farther southeast faced the Lorraine plateau, fade away in this northern district; hence here, after the repulse of the allied forces in northern France in August, 1914, the German army, hurrying through the Ardennes highlands by the gorge of the Meuse and passing around the low western descent of the highlands by Charleroi and Mons in Belgium, had a wide district of moderate relief before them, across which they advanced rapidly far southward into the Champagne, and for a less distance over the high ground of the first upland belt, to which we now turn.

CHAPTER IX

THE FIRST UPLAND BELT

47. *The Scarp of the Upland facing the Champagne Lowland.* The member of the series of upland belts nearest Paris is maintained by beds of impure limestone overlying the weaker sands and clays, which are worn down in the lowland of the Champagne on the east. Through the middle of its arc the upland has a strongly developed east-facing scarp, 150 m. or more in height, a rather broad upland (240–280 m.), and a long westward slope toward Paris. Its beginning may be said to be in the uplands of the Forest of Fontainebleau, not far southwest of the transverse valley of the Seine. The upland gains stronger relief to the northeast of this valley, where it is known as *la Brie*, as far as the deeper and more meandering valley of the Marne, in the open entrance to which lies Epernay, with Chateau-Thierry in its narrowed middle part, and Meaux near its exit to the lower ground near Paris. The main line of the Eastern railway follows this valley. The lowland on the west has been described in chapter III.

The upland of *la Brie* is cut through about two-thirds of the way from the Seine to the Marne by the Petit Morin, the source of which lies in an extensive marsh, *le Marais de St. Gond*, on the lowland next east of the upland scarp. Several streams, of which the largest is the Grand Morin, a short distance south of its more deeply incised little brother, rise on the upland and cut valleys in its back slope. Beneath a notch in the scarp at the head of the Grand Morin lies

Sézanne on the adjoining lowland; a secondary line of the Eastern railway ascends the Grand Morin valley and passes through the notch to the lowland. It was on the slanting upland just north of the Petit Morin valley, that Napoleon, shortly before his exile to Elba in 1814, defeated the Prussians in three battles at Montmérail, Champaubert, and Eloges on three successive days.

The escarpment of the Brie upland is known as the *Falaise* (sea cliff) *de l'Ile de France* (*Ile de France* being the name of an ancient province, centering in Paris), thus recalling the obsolete view that cliffs of this kind were, like the cliffs of Normandy, the work of sea waves, as was thought before an understanding was reached of their origin by the differential weathering of strong and weak strata. The fact is simply that the underlying weak strata, worn down on the east to lowland forms, slowly sap and force back the scarp of the harder overlying strata; while the overlying strata, resisting with all their strength the enforced retreat, stand forth in scarped promontories and spurs. All the sunlit slopes of the scarp and the northern side of the Marne valley are occupied by vineyards from which the famous *vin de Champagne* is produced; it is stored in great cellars excavated in the weaker sandy strata near the scarp base.

The finest promontory of the upland scarp is the already mentioned strong salient, the *Montagne de Reims*, next north of the Marne; its eastward-rising extremity (280 m.) is naturally the highest point in the whole length of the scarp; its forested back slope is drained northwestward by a little stream, the Ardre, to the Vesle, a tributary of the Aisne. The promontory is tunneled by the railway from Epernay to Rheims.

The segment of the upland front, almost detached by the valley of the Ardre, has an irregular margin, thus foreshadowing the still more irregular margins of the segments north of the Vesle and the Aisne.

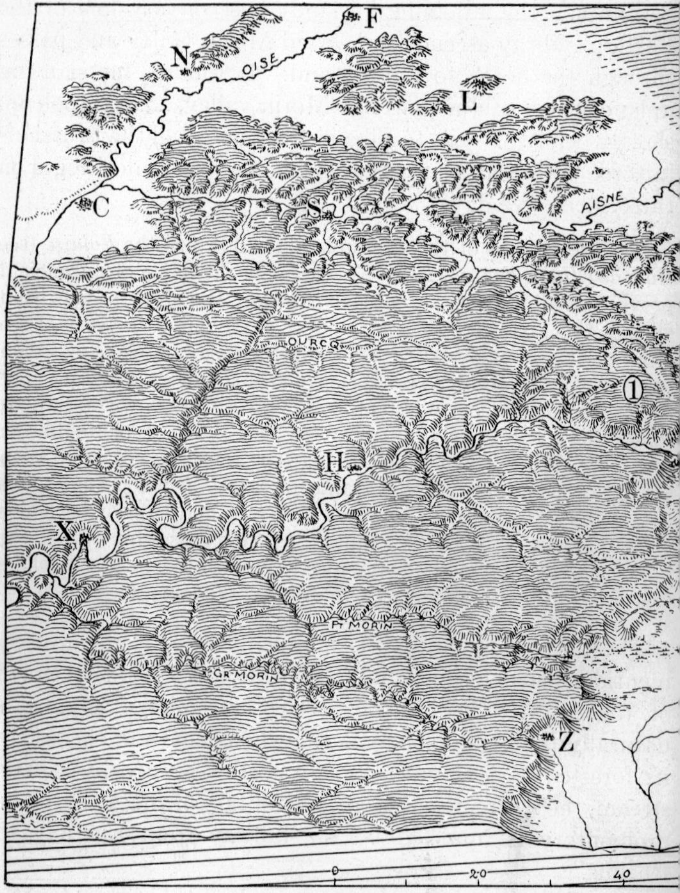

FIG. 34. THE COURSES OF THE MARNE AND THE AISNE ACRO

Explanation of abbreviations: A, Châlons-sur-Marne; B, Bar-le-Duc; C, Compiègne; E, Epernay; F, La Fère; H, Château-Thierry; L, Laon; M, Ste. Menehould;

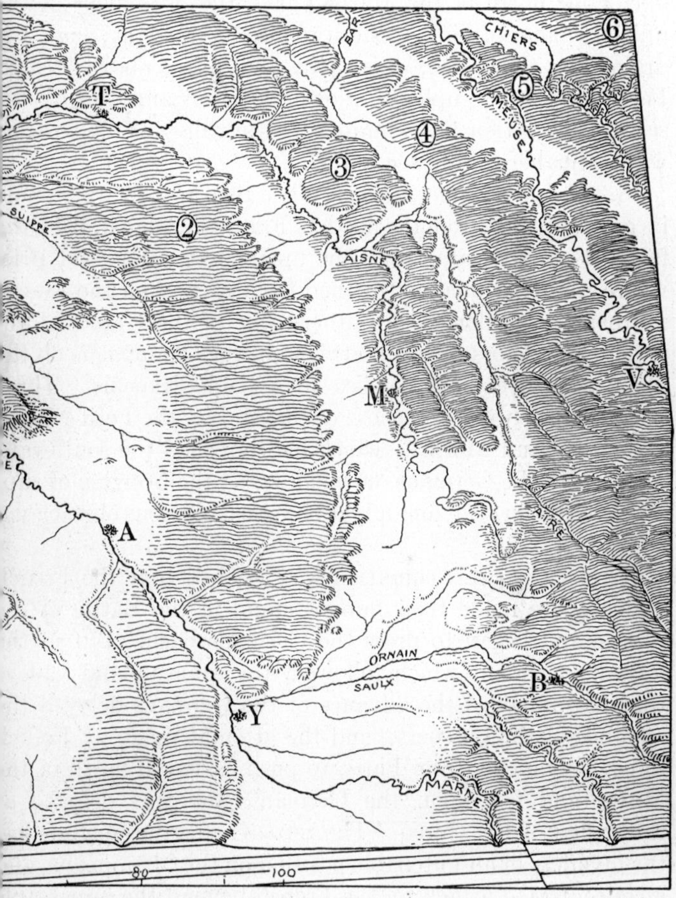

ɪᴇ Cʜᴀᴍᴘᴀɢɴᴇ ᴀɴᴅ ᴛʜʀᴏᴜɢʜ ᴛʜᴇ Uᴘʟᴀɴᴅs Nᴏʀᴛʜᴇᴀsᴛ ᴏғ Pᴀʀɪs

N, Noyon; O, Fère Champenoise; R, Rheims; S, Soissons; T, Rethel; V, Verdun; X, Meaux; Y, Vitry-le-François; Z, Sézanne.

Not far north of the Marne valley the back slope of the upland changes from a western to a southern slant; here it is drained by the Ourcq, which joins the Marne above Meaux. Farther west, the upland, known as Valois and already described in section 17, is limited by the valley of the Oise, as will be further told in section 52.

48. *The Battle of the Marne.* When the German army made its great advance from the north of France in August, 1914, the farthest progress was over the uplands of Valois and Brie where the front finally stretched from Meaux on the Marne southeastward along the southern side of the Grand Morin valley, and across the open lowland of the Champagne on the line from Fère Champenoise to Vitry-le-François. Three valleys in the upland—those of the Marne, the Petit Morin and the Grand Morin — which had impeded the southward progress of the Germans, now hampered the bringing of supplies from their rear, and might become dangerous obstacles in a forced retreat.

It was when this condition was reached that the French under Joffre made their famous stand in the Battle of the Marne and began to press the Germans back. But in the forced retirement which followed, the same upland valleys that had impeded the advance of the Germans delayed the advance of the pursuers; and the attempt made by French forces, advancing from Paris, to push eastward north of the Marne and fall upon the German flank was delayed at the valley of the Ourcq. The retreat over the uplands left the invaders in an untenable position on the Champagne, and a concentrated attack by the French beyond the impassable marsh of St. Gond, which served as a natural barrier for a moderate distance forward from the upland scarp, compelled the German army on the plain to withdraw from its hazardous isolation.

The northern side of the Marne valley would have been a favorable line for the Germans to hold, had there been time to assemble their forces upon it, but the pursuit was so ardent that this proved impossible; and moreover the right flank was there exposed to attack on the uplands of Valois. The retreat was therefore not arrested until the valley of the Aisne, in the uplands farther north, and the hills on the Champagne east of Rheims were reached. The positions there taken were well chosen, if one may judge by the long subsequent period during which they were held with small change.

49. *The Tablelands north of the Aisne.* The northern part of the first upland belt is cut through by the east–west valley of the Aisne, midway in which Soissons is situated. The cuesta-like upland here undergoes the change of form that such features suffer, as illustrated on page 48, when the dip of the determining strata decreases and becomes almost horizontal. The upland belt or cuesta, with a scarp on one side and a long slope on the other, thus becomes a plateau, with a flat upland surface and scarps on all sides, as shown in the bird's-eye diagram, pp. 116, 117. The change is foreshadowed in the plateau-like segment (200 m., see Fig. 34) of the gently slanting upland included, with scarped and indented margins on all sides, between the Vesle and the Aisne, northwest of Rheims; and also farther west, on the south side of the Aisne valley in the neighborhood of Soissons, where the upland margin (140 m.) has lost the comparatively direct course that characterizes well-defined upland belts, and become irregularly lobate, after the fashion of dissected plateaus. North of the Aisne valley the transformation is complete; here instead of

presenting a frontal scarp, an upland, and a back slope, the continuation of the first upland is now seen in several separate plateau segments varying little in altitude (140–180 m.),

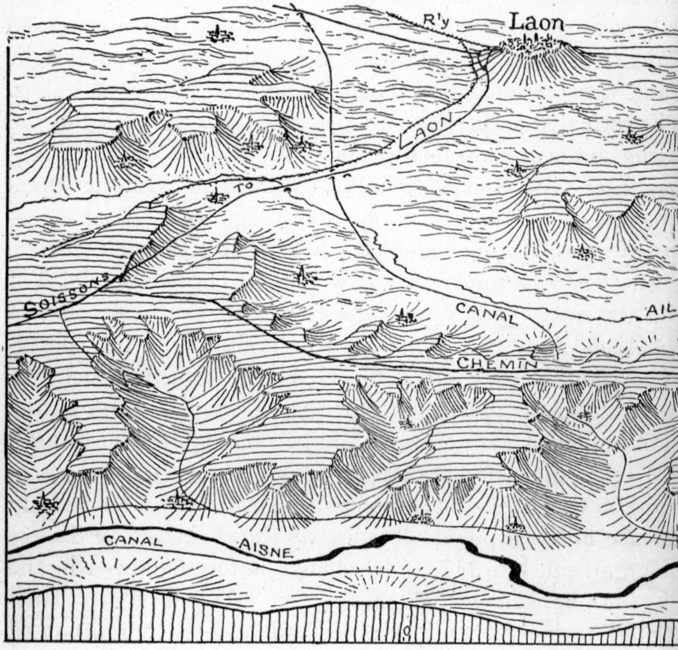

FIG. 35. DIAGRAM OF THE VALLEY OF T

with strongly indented margins that fall off on all sides in steep scarps and long lower slopes.

The first of these segments stands between the nearly parallel valleys of the Aisne and the Ailette, a small branch of the Oise; it measures 60 k. in length east and west, and near its western end is about 12 k. in width; it narrows eastward and Craonne lies on the slope below its slender extremity. A canal passes under a narrowed part of the plateau by a tunnel excavated in the weak strata underlying the lime-

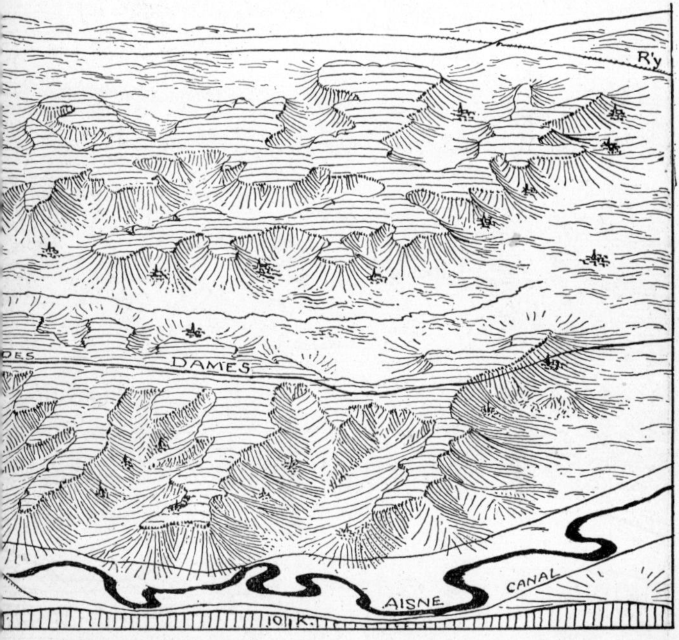

ISNE, THE CHEMIN DES DAMES, AND LAON

stone beds that form the upland, and thus connects the Aisne and the upper Oise. The Oise flows south beyond the broader western end of the plateau segment; Compiègne lies on the east side of this river near the confluence of the Aisne. The margin of the long plateau segment or tableland is frayed out, especially on the southern side, in many spine-like spurs

FIG. 36. THE VALLEY OF THE AISNE, AND

...EAU SEGMENT BEARING THE CHEMIN DES DAMES

between as many encroaching, roundheaded reëntrants. A road running northward from Soissons to Laon crosses over the narrowed middle of the plateau segment between two reëntrants; a branch road runs along the level upland in the eastern half of the segment, avoiding the reëntrants on either side: this is the famous *Chemin des Dames*, shown in Fig. 35 and in more detail in Fig. 36. It was in this district that Napoleon was at last overcome in 1814, shortly after his three victories on the upland over the Petit Morin; the downfall of the First Empire and his exile to Elba followed.

North of the Ailette valley, the breadth of which is much greater than might be expected for so small a stream, two smaller plateau segments or little tablelands are separated by an open depression; they have elaborately frayed-out margins; the western and larger one (220 m.) is mostly covered by the forests of St. Gobain and Coucy (Fig. 34). Next north of the depression between the two isolated segments is a small outlying hill, crowned with the walled city of Laon. Two more small detachments of the upland rise west of the Oise; Noyon lies between them. North of Laon, the extension of the northernmost sector of the broad Champagne belt stretches westward beyond the Oise, as will be further described in section 52.

In this district, much more easily than farther south where the first upland belt is developed as a cuesta, one may recognize that the controlling strata, now divided into detached plateau segments, were originally continuous and that the strata then extended with slowly increasing altitude far eastward and northward over the adjoining lowland; one may, indeed, if he pause here long enough to consider the relation of the brief present and the long past, come to understand that the removal of the extended strata and the disseverment of their remaining plateau segments was accomplished, not by any convulsion of nature, nor yet by the hurried processes of overwhelming floods, but by the quiet perseverance of the slow processes of weathering and washing, working unendingly just as they are working today: they include the incision of valleys by streams, the gradual disintegration of the plateau-making strata exposed on the valley sides, the unceasing but very deliberate creep and the occasional more rapid wash of the disintegrated detritus down hill to the streams, and its intermittent transportation along their channels to the sea.

The inevitable result of these processes is the slow reduction of the plateau segments to less and less area, and the corresponding increase in the breadth of the intervening valleys. South of Soissons the valleys are comparatively narrow, and the plateau areas are large; the valleys of the Aisne and the Ailette are more broadly opened, and the plateau segments are small and isolated tablelands;

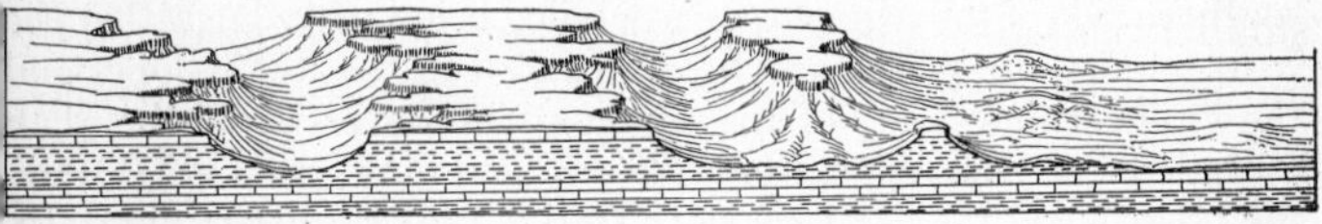

FIG. 37. WASTING TABLELANDS REDUCED TO LOWLANDS

farther north, the valleys become wide lowlands, and the tablelands are reduced to still smaller dimensions, like the hill on which Laon stands; farther north still, where the plateau-making strata rose gradually to greater altitudes and were therefore more exposed to degradation by the sapping of the weak underlying strata, no tablelands now remain and the lowland is continuous.

50. *Contrasts of Upland Belts and Tablelands.* A difference between upland belts or cuestas of gently inclined strata and plateau segments of horizontal strata, of importance with respect to the excavation of trenches in a "war of positions," is seen in the unsymmetrical cross-section of cuestas, Fig. 14, and the symmetrical cross-sections of plateau segments, Fig. 37; or to put it in another way, in the unsymmetrical valleys that separate adjacent cuesta-uplands, and the symmetrical valleys that separate adjacent plateau segments or tablelands. The two sides of an inter-cuesta valley or depression are of unlike slopes and are underlaid with unlike strata: an example of this kind already treated is the depression between the back

slope of the sandstones that maintain the third upland belt in the Forest of Argonne on the east and the low frontal scarp of the second or chalk upland belt on the west.

On the other hand, the Aisne valley, between the long plateau segment that carries the Chemin des Dames on the north and the ragged border of the Valois upland on the south, is of the same structure on its two sides. The battle front lay in this valley for nearly three years, and while the northern side was in possession of the Germans, the French on the southern side could learn from experience in their own trenches and dugouts all the peculiarities of the enemy's position, so far as they were dependent on rock structure. Likewise the two sides of the long plateau segment between the valleys of the Aisne and the Ailette are of the same structure: hence, when the French, in the spring of 1917, drove the Germans up the southern side of the segment, as is detailed below, they knew that the weak sandstones which had been trenched in the lower slopes and the stronger limestones of the capping bluffs over which they had ascended from the Aisne, would be repeated in reversed order when they descended the northern slope to the Ailette, as they did a few months later; and that the same features would be encountered again when the smaller segments, lying farther north, come to be attacked.

51. *The Aisne Front.* The position taken by the Germans after the battle of the Marne ran from the southern part of the forest of Argonne—the third upland belt—obliquely over the second or chalk upland belt and across the Champagne lowland northeast of Rheims to the Aisne; then down the valley of the Aisne almost to Soissons, where the line crossed obliquely over the plateau segment next north and traversed the Oise between Compiègne and Noyon. The French captured the middle part of the slope on the south side of the plateau segment near Soissons in September, 1914, but could not hold it and were forced back across the river. Not until two and a half years later was the German front along the Aisne pushed definitively northward. In March, 1917, when an important retirement of the Germans was made farther

north (see section 54), they abandoned the broad western end of the long plateau segment and withdrew up the Oise; Noyon was thus liberated, and Soissons, which had for more than two years suffered bombardment, enjoyed comparative quiet again and the restoration of train service to Paris. A month later, on April 16 and 17, an important movement was made along the valley of the Aisne east of the confluence of the Vesle: the French had been entrenched for about 30 months on the lower slopes of the upland next south of the river, while the Germans had occupied the slopes at the base of the long plateau segment next north of the river, through the villages of Vailly and Chavonne, shown on the map, pp. 118, 119.

The Germans were sheltered in well prepared dugouts during a preliminary bombardment by the French; when the shells ceased falling and the hidden Germans emerged with their machine guns to repel the expected assault, the up-hill charge of the French was so rapid and vigorous that thousands of the trench defenders hastily surrendered, and the reserves hurriedly fled to the upland crest. The village of Craonne, on the slope below the limestone bluff at the slender eastern extremity of the plateau segment, was captured in ruins on May 15; the narrow terminal part of the segment, curiously enough called *le plateau de Californie*, was taken the next day as well as a stretch of the upland farther west, where the road from Soissons to Laon crosses it and where Laffaux mill stands west of the fork of the Chemin des Dames; on the following days greater lengths of the Chemin des Dames were gained: but so stubborn was the resistance on the plateau that several months were there spent in attacks and counterattacks and not until the autumn of 1917 were the Germans forced to retreat down the northern slope and retire across the broad valley floor of the Ailette toward Laon.

CHAPTER X

THE REGION BETWEEN THE UPPER OISE AND THE SOMME

52. *General Features of the Region.* If a radius trending northeastward from Paris through the upper Oise be turned until it runs northward through Amiens on the Somme, the 100 k. of its outer length, between 70 and 170 k. from the Paris center, will sweep over a sector of rolling surface, mapped on p. 134, partly a lowland of sands and clays (80 to 100 m.) partly an upland of chalk (120 to 160 m.) which constitutes the western continuation of the lowland and upland sector already described between the Aisne and the Oise. It is abundantly incised by open and irregularly branching valleys from 25 to 50 m. in depth, shown in detail in Fig. 38. The region therefore has what may be called a quilted surface, inasmuch as the inter-valley areas, which rise in broadly rounded hills, are rather uniformly convex, so that when many of them are viewed from the center of any one, they all unite in a nearly level skyline; while the valleys into which the convex areas gradually descend are comparatively narrow, like the seams by which a quilt is furrowed. Only the larger streams have well developed flood plains, usually less than a kilometer across; their valleys are like double seams.

The highest elevations commonly occur along the broadly arched divides between the headwaters of neighboring river systems, of which the most important separates the Escaut on the north and the Oise and Somme on the south, and trends roughly east and west. The elevations along the divides are too low and too rambling properly to deserve the name of ridge, but as our language has no better word for them, that name is commonly employed. Distinguishing landmarks are almost wanting; the landscape repeats itself with small change from place to place. It is only in the southeastern part of the lowland area (80 or 90 m.) west of the middle Oise next above the confluence of the Aisne, that tabular hills surmount the lowland; they are small outlying segments of the plateau of the Aisne-Ailette area and are capped with the same limestone (*calcaire grossier*); similar but larger tabular hills occur farther down the Oise, as described in section 18. Elsewhere the capping limestones, which once spread far over the region, are completely removed; even the next lower strata of sands and clays have but a moderate extension west of the Oise, and beyond them the still lower flint-bearing chalk formation is laid bare at the beginning of its wide extension in northwestern France.

Conformably to the general structure of the Paris basin, the more resistant chalk strata hereabouts rise very slowly to the northeast, north, and northwest from beneath the weaker sandy and clayey strata, and thereupon the lowlands are succeeded by slowly rising uplands. Chalk is a rock of moderate hardness, easily pervious to water; hence deep wells are needed on the uplands to reach a water supply. The soil resulting from its decomposition is a reddish clay, usually of moderate depth, charged with flints; it represents the insoluble parts of the rock after the chalk is removed by solution. The clay soil is peculiar having sudden variations of depth, as if the chalk were more soluble or solution had been more active at one point than at another. It is, however, chiefly on the slopes and valley sides that the clay soil forms the surface; the broad hill crests are usually coated over with a fine and fertile loam, easily trenched, and very quickly converted into mud in wet weather.

Woodland areas, comparatively small hereabouts, are now desolated. The main roads, formerly bordered by evenly

FIG. 38. THE UPLAND ABOUT

NTIN AND THE UPPER OISE

spaced trees, often having closely trimmed trunks and a tuft of branches at the top, are now bare. As elsewhere in France, pastures and cultivated fields, abutting directly upon the roads, occupy nearly all the surface; the open landscape is therefore characterized by many straight-line strips and patches, varying delicately in color according to their crops. Little or no space is given to hedges, walls, or fences; formerly, grazing cattle were picketed and flocks of sheep were restrained from wandering by shepherd dogs. But now, as around Verdun, the country is laid waste; the digging of trenches and the blasting of shell-craters has done a lasting injury to the fields by mixing the humus-bearing surface soil with the subsoil and the underlying rock; occasional unexploded shells, buried in the ground, may make plowing dangerous. It will be many years before the fertility of these devastated areas is fully regained.

The relief of the surface is generally so moderate that the main roads not infrequently run on direct courses for distances of 5 or 10 k. Some of them follow ancient Roman roads, and on these the long established villages are not infrequently located; but the *routes nationales* of modern construction are commonly laid on lines that serve best to connect the distant larger towns and cities, and therefore leave many villages to one side or the other without turning to enter them; these villages are therefore served by local roads. Hence troops advancing along the old Roman roads find more frequent shelter in villages than if the advance is made along the new *routes*. The railways, having to respond more closely to the form of the surface, are somewhat sinuous.

53. *Rivers and Cities.* The chief streams of the region are, in the north, the headwaters of the Escaut, and of its western branches, the Scarpe and the Sensée; on the west the upper Somme. The Oise forming the eastern border of the region, and the Sambre farther north, receive no important tributaries

from the area here considered. The larger streams wander in underfit fashion on flat and marshy flood plains, from a half to one and a half kilometers in width; for as already explained in the case of the Meuse a considerable part of the drainage creeps slowly through the alluvium of the flood plain as an "underflow," and the visible river therefore does not represent all the run-off of the rainfall. Below Amiens and therefore outside of the area here considered, the Somme valley is unusually rectilinear; above Amiens it is remarkably sinuous for 25 k.; here the small river wanders very irregularly as an underfit stream on the low and marshy valley floor. The upper Oise has a similarly irregular course through its flood plain, but its valley is relatively straight, as shown on the detailed map, page 127. The other valleys do not present peculiar features.

Canals, almost as numerous as the larger streams, serve to connect the industrial region of the lowlands farther north, where coal mines and factories abound, with the metropolis on the Seine system to the south; the canals not infrequently tunnel under the divides; a tunnel of this kind five k. in length, north of St. Quentin, passes under the divide (142 m.) between the headwaters of the Somme and the Escaut.

Excepting Amiens (93,207) on the Somme, there are no large cities in this region; but the names of many smaller cities have become famous during the War. Arras (see Fig. 40) lies on the Scarpe near the descent of the chalk uplands to the lowlands of northernmost France or French Flanders, to be described in section 60; Péronne and St. Quentin lie on the uppermost Somme; Albert and Roye are on its branches; Noyon stands between two of the tabular hills west of the middle Oise; La Fère and Chauny lie in the Oise valley where it approaches the tabular hills north of the Aisne. The villages of the region are arbitrarily placed on the inter-valley arches, on the slopes or in the valleys.

54. *The War Front from the Oise to the Scarpe.* Interest in the simple landscapes of this region centers at present in their relation to the shifts of the fighting front between the Allied and the German armies. From September, 1914, to the summer of 1916, the line of contact suffered little change; trending southward from the Belgian frontier near the coast across the lowlands of French Flanders, it ascended to the chalk uplands near Arras and continued to Noyon, where it turned eastward across the middle Oise to the valley of the Aisne, as noted above. Midway between Arras and Noyon, the line crossed the sinuous valley of the Somme, and there in early July, 1916, the battle of the Somme resulted in an advance toward Péronne. Three months later the advance was increased, and the indentation thus made in the German line endangered the adjacent salients, one on the north including Bapaume between the Somme and the Scarpe, the other of larger area on the south including Roye and Noyon between the Somme and the Oise.

The vertex of the Bapaume salient at Thiepval was cut off in mid-November, 1916, and at the end of that year, Péronne was almost reached. A still closer approach to Bapaume was made near the close of February following. Thereupon the Germans made a "strategic retirement," barbarously devastating the countryside as they withdrew, so that in March, 1917, the two salients were reduced to a rectified front, running in a comparatively direct line from near Arras, south-southeast to La Fère on the Oise, where the valley plain was flooded by opening the canal further north as a means of protecting the retreat; thence over the tabular hills to the Aisne below Soissons. Thus Bapaume in the northern salient,

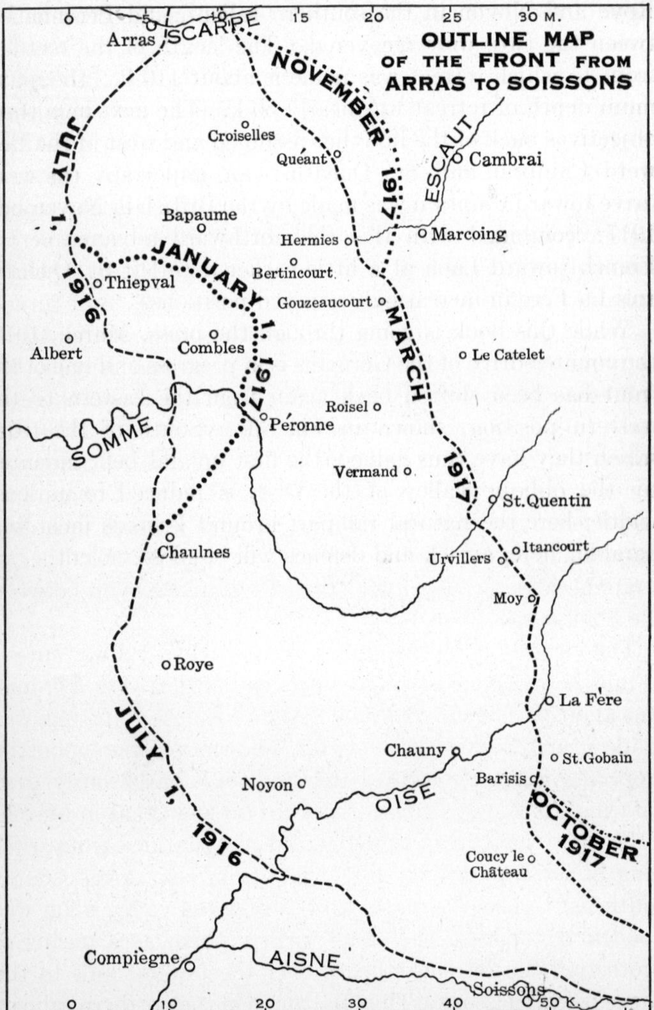

FIG. 39. MAP OF THE FRONT FROM ARRAS TO SOISSONS

Roye and Noyon in the southern salient, and Péronne between the two, were recovered. The length of the rectified front to which retreat was made is about 110 k.; the maximum depth of retreat was nearly 50 k. The next important objectives back of the line thus assumed and west of the Oise were Cambrai and St. Quentin: an important eastward drive toward Cambrai was made by the British in November, 1917; combined with this the northward advance of the French toward Laon at a little earlier date left St. Quentin and La Fère in increasingly exposed positions.

While this book is going through the press, March, 1918, the counter-drive of the Germans is in progress and line of the front has been shifted back again from the eastern to the western position, shown in Fig. 39. South of the front which they have thus gained, the first upland belt, indented by the oblique valley of the Oise, is reduced to its least width; here the natural rampart around Paris is most vulnerable: here, attack and defence will be most violent.

CHAPTER XI

THE NORTHWESTERN UPLANDS

55. *The Chalk Country of Picardy and Normandy.* The wide saddle of nearly horizontal chalk strata, between the Ardennes highlands on the northeast and the Armorican hills beyond the Seine on the southwest, includes the uplands of the old provinces of Picardy and Normandy. These uplands reach altitudes of from 130 to 200 m., and occupy a belt which borders the English Channel for 220 k. from Cape Gris Nez to the Bay of the Seine and extends 80 k. or more inland; they are adjoined on the northeast by the lowlands of northernmost France, described in the following chapter; they decline gradually southeastward to the lower land along the middle Oise; farther south toward Paris they are overlapped by the uplands of the Vexin salients, described in section 18. The underlying chalk controls the features of all this region: indeed, an area stretching northward from the lower Seine to the coast is known as the *Pays de Caux* or the Chalk-land (*Caux*, provincial for *Chaux* = Latin, *calx;* English, *chalk*). The features here encountered therefore repeat those described in the preceding chapter, but with somewhat greater relief.

The uplands, divided by irregularly branching valleys, undulate gently in broad, faint arches one after another, and are of uniform appearance over large areas. Their surface is usually cloaked with a fine and fertile loam; where this is absent, as is commonly the case on the slopes toward the

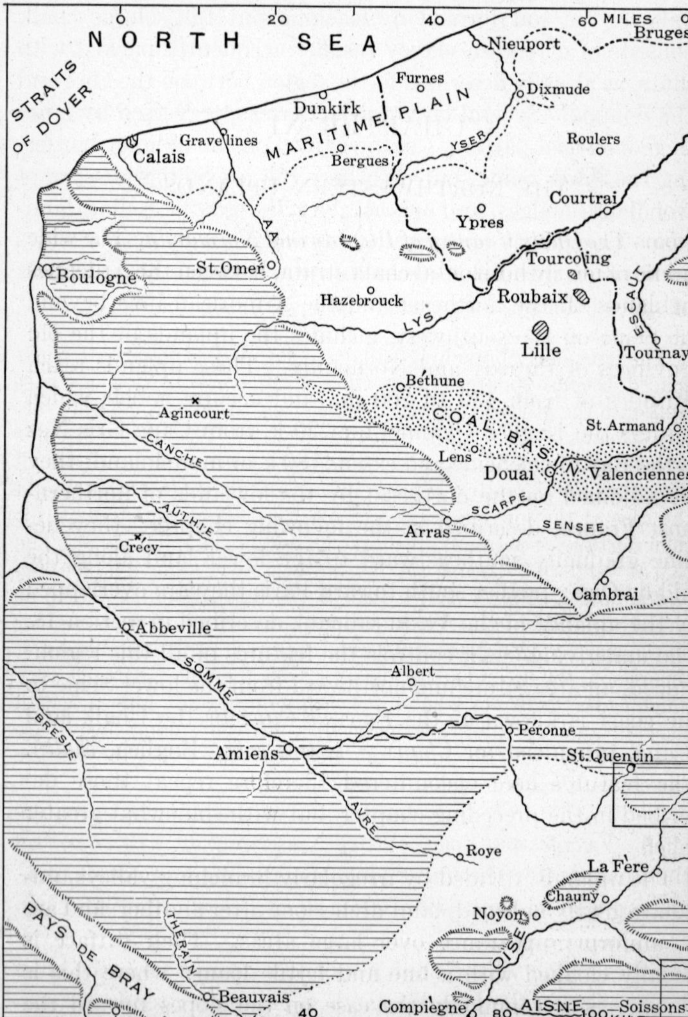

FIG. 40. THE UPLANDS OF NORTHERN FRANCE

valleys, the soil formed by leaching out the soluble chalk consists of a reddish clayey residue, frequently packed with flints, as already described in the region between the Oise and the Somme. Several large forested areas, traversed by lanes in geometrical patterns, are preserved on the uplands bordering the lower Seine, but elsewhere the prevailing absence of woodland, hedges, and fences gives the gently rolling landscape a peculiarly open quality over large areas, in spite of its subdivision into fields, crossed in the distance by road-lines of spaced trees.

The irregularly branching valleys of the chalk uplands are wide-spaced, because so large a proportion of rain water has for ages past sunk into the ground that little has been left to act as wandering surface streams in the work of valley excavation. The smaller streams extend, as commonly happens in areas of horizontal strata, in all directions indifferently; but exception must be made to this rule in the case of the lower Somme and several neighboring streams, for their northwestern trends are markedly rectilinear and parallel. The rule of wandering streams obtains, however, over a much larger area and is nowhere better shown than in the upper Somme, which, as already described in section 53, winds about in an uncertain manner on the marshy flood plain of its serpentine valley. Through most of this region, the upland spurs interlock with the valley bottoms very much as the fingers of a hand, held palm down, may interlock with those of another hand, held palm up; but the upland fingers are broad with irregularly lobed margins, and the valley fingers are narrow with many irregular branches.

56. *Villages and Roads.* Villages are compactly clustered on the uplands, for the underlying chalk being a pervious rock, ground water is sometimes 60 or 80 m. below the upland surface; hence economy demands that one well shall serve many families. The village churches or tree-groups form the chief landmarks on the broad uplands. The better roads are sur-

faced with broken flints packed in chalk, and are gleaming white in sunshine; they serve well if well used, but the flints present many sharp edges if ordinary traffic is so infrequent as to let rain-water wash off the binding material or if heavy traffic wears it away.

Some of the roads and railways follow relatively direct courses across the uplands, rising and falling with the gentle undulations of the surface, turning slightly to approach a larger village or curving around a valley head to avoid steep descents or to save the heavy cost of viaduct construction; others follow the larger valleys, assured of easy and steady grades, but turning as the valleys turn and bridging the main and side streams as occasion requires; still others connect the upland and the valley roads by branching along little side-valleys that unite the two levels. The main line of the Northern Railway — le Chemin de Fer du Nord — from Paris to Calais, via Amiens, shows many of these devices.

Woodland is more common on the valley sides than on the uplands. The higher parts of the valley ends are dry; running streams are not found until the valley is followed down to a considerable depth below the upland surface; there the verdant valley floors are wide enough for small fields and gardens, near which every house may have its own well, for ground water is here found at little depth; hence the compact villages around the deep wells of the uplands are replaced in the valleys by long villages of detached houses.

Some of the larger valleys, like that of the Somme, have wet and marshy flood plains in which the water area is increased by the excavation of peat for fuel and by the construction of fish ponds. The lower valley of the Somme is also notable for its gravel terraces of moderate height but of great scientific interest, for it was in these terraces near Abbeville, 25 k. inland from the coast, that to the astonishment not to say incredulity of the world Boucher de Perthes seventy-five years ago found artificially chipped flints, the first authentic traces of prehistoric man, the contemporary of animals now extinct, and thus made a beginning for the science of archaeology.

57. *Valley of the Seine.* The lower Seine is the chief river of this region; it has so winding a course that a direct distance

of 70 k. inland from the head of its estuary is nearly doubled if the river is followed. Each convex curve of the winding river sweeps around the base of a great amphitheater, 3 to 5 k. in radius, alternately on the right and left bank, where the upland is undercut to a steep descent; and into each river curve, alternately from the left and right, a spur slopes down from the opposite upland. Scroll-shaped belts of flood plain, sweeping around the spur ends, border the river, now on one bank, now on the other, in the manner described for the looped course of the river below Paris in section 19.

The tide, advancing through the narrowing estuary of the Seine, is crowded into a rapid current which at certain times each month rises rapidly (4 meters in 30 minutes) and rushes 50 or 60 k. up the river, assuming for part of this distance the form of foaming waves, known as *le mascaret*, and causing a quick rise from low to high water; the flood current may be so swift that river boats omit landings while it is running. The fall from high water to low is more leisurely; then the normal down-valley current prevails. The rapid inflow must have been an aid to the sea-faring Northmen, when they came up the river in the ninth century in their boats as far as Rouen and established themselves in the district which, as Normandy, has since borne their name. The tidal currents give so great a reënforcement to the normal current of the river through the lower 30 k. of its estuarine length, that the upland spurs, around which the river farther upstream flows in so winding a course, are here nearly worn away, and the deep-water channel through the low-tide mud flats of the widened estuary is almost directly from east to west.

The buoys along the channel must be carefully regarded. It is recorded that a British steamer some years ago unwarily ran half aground on the channel border at high tide; the ebb current scoured away the mud under the bow; bow and stern being then unsupported at low water, the vessel broke in half and the two ends tilted down, leaving the broken hull high and dry amidships; the next flood and ebb currents scoured away the mud under the break, and at the second low water, both parts of the hull settled deeper than before; after a few days the vessel was almost buried out of sight, a total loss.

The modern commercial city of Havre (136,159) lies on the northern shore of the estuary. At the third pronounced northeastern curve of the Seine counting up stream, and 70 k. in a direct line east of Havre, but 120 k. following the river, the ancient city of Rouen (124,987) is situated on the right bank: here are the first bridges over the river. A spur of the upland next east of the city, 160 m. in altitude, is ascended by an inclined railway and offers a fine prospect over the open valley and the enclosing uplands. Elboeuf is on the next southwestern curve of the river above Rouen. Navigation up stream to Paris is made possible for good sized vessels by locks. An important line of the Western Railway — Chemin de Fer de l'Ouest — follows the winding valley around its larger curves from Paris to Rouen, but shortens its route somewhat by crossing the river several times and by tunneling through the narrower spurs; its continuation from Rouen to Havre passes over the uplands.

No cities of importance are found on the inter-valley uplands; they are all in the larger valleys or on the coast. Beauvais and Neufchâtel lie in or near the depression of the Pays de Bray (described in the second following paragraph), Amiens and Abbeville are on the Somme; while Fécamp, Dieppe, and Boulogne-sur-Mer are on the coast, as will be further stated below. Some of the upland villages are of historic interest as the scene of battles between the French and English, during the period in which the Plantagenet kings of England tried to maintain territory in France inherited from their Norman predecessors; thus Crécy where Edward III defeated the French in 1346 lies on the upland north of Abbeville; Agincourt where Henry V defeated the French in 1415 is some 30 k. farther northeast (see map, page 134); but forty years later, the French drove the English out of France.

58. *Exceptional Features.* A district of exceptional features, known as *le Boulonnais*, is found in Artois, the northernmost province of France; it borders the narrowest part of the Channel opposite Dover, where the coast makes out in the salient of which Cape Gris

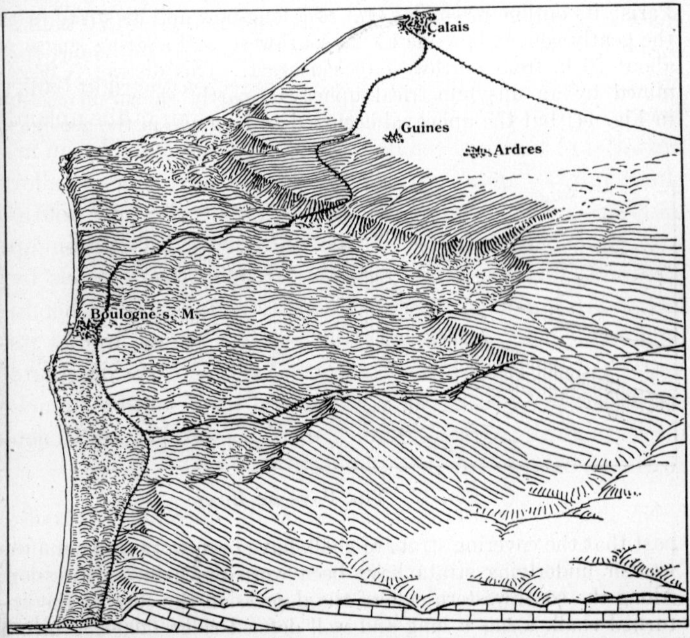

FIG. 41. THE BOULONNAIS, BETWEEN BOULOGNE AND CALAIS

Nez is the extremity, with Calais on the northeast and Boulogne-sur-Mer on the south. A broad and ancient domelike upheaval of the strata extends across the Channel from southeastern England and here enters France for 25 k.; as a result the chalk beds, which elsewhere cover the uplands, have been worn away from the upheaved area, so that their slanting edge forms an irregular rampart around an enclosed space in which the underlying strata, eroded into hills, meet the coast in cliffs of variable structure and form. It was in the gently slanting strata on the outer slope of this upheaved area that the first "artesian" well was bored.

Another exceptional district is a lowland belt, known as the *Pays de Bray*, which crosses the middle of the chalk uplands northwest of Paris; its outline resembles that of a long-bow and its string, with the gently convex bowside to the southwest, and a string length of about 70 k. from northwest to southeast. This district is determined by an unsymmetrical upheaval, partly shown in section in Fig. 44; but the upheaval occurred so long ago in the geological

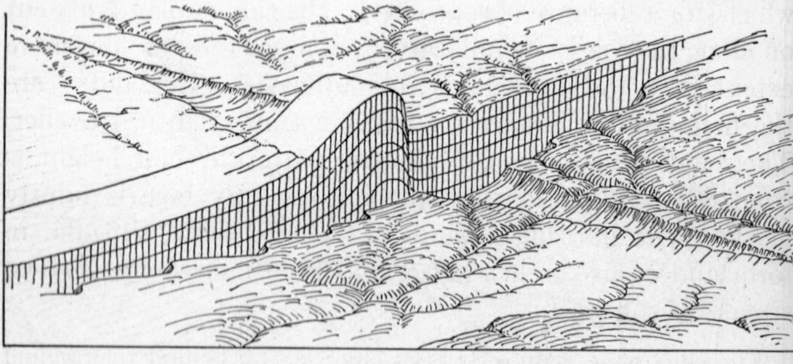

FIG. 42. THE PAYS DE BRAY

past that the covering strata of chalk have been worn away and the weaker underlying strata have been worn down in a depression. Along the southwestern side of the depression the edge of the upturned chalk forms a long and well defined ridge (200–230 m.), a superb line of defence from attack on the north.

The Pays de Bray is therefore a good example of a lowland that has been eroded along an upheaved belt of weak strata, and that is enclosed by uplands of less upheaved but more resistant strata. The lowland is partly drained northwestward to the coast by the rectilinear Bethune, at the mouth of which lies Dieppe; but also by the Andelle and the Epte which flow from the central part of the lowland southward through the chalk uplands to the Seine. The Thérain flows southeastward to the Oise in a valley of the uplands next northeast of the lowland. The fertile soils of the lowland and its branch valleys are noted for their dairy farms. Its chief towns are, as above noted, Neufchâtel near the northwestern end and Beauvais on the northeastern margin. It is the erosion following the Pays de Bray upheaval, where it was prolonged with decreasing strength to the southeast across the Oise, that causes the separation of the smaller upland salients west of the middle Oise below Compiègne from the larger upland salient that extends west of the lower Oise, as described in section 18.

59. *The Clift Coast along the Channel.* The sea cliffs by which, for a distance of some 200 k., the chalk uplands are cut off along the coast — similar to but of much greater 'longshore extent than those of Dover in southeastern England — are the first sight of France for many a transatlantic traveller. Their outline is of gently sinuous pattern; their height is from 80 to 130 m.; their precipitous gray face is faintly marked with horizontal bands of included flints, irregular in form and of fist or head size, and is disfigured by vertical streaks of soil-wash.

The cliffs have manifestly been cut back far behind the original seaward extension of the chalk strata. Little valleys in the uplands are sharply cut off in the cliff face. They sometimes converge toward the shore line, as if to join a trunk valley farther on; but the trunk valley as well as the upland which enclosed it have been undercut and consumed, and their space is now occupied by the sea. If one walks along the cliff top, whence a beautiful view over the Channel may be enjoyed, fissures may be seen parallel to the cliff face and a few feet back from it, presaging the detachment and fall of a great slab or block; the older furrows of a plowed upland field may sometimes be noted on the dangerous side of the fissure; detached slabs, shattered by their fall, are occasionally to be recognized in heaps of fragments at the cliff base, not yet wholly removed by the restless sea.

A sheet of beach detritus spreads seaward from the cliff base on the slowly deepening bottom; it consists of cobbles and pebbles on shore, and of pebbles and sand farther off shore; the cobbles, pebbles, and sand are the waterworn fragments of flints which withstand the beating of the waves long after the chalk in which they were embedded in the cliff face has been disintegrated and dissolved to a residual silt and washed far out to the deeper sea floor. The vertical range of the tide from high to low water is six or seven m.; the flood and ebb 'longshore currents are strong. The beach is submerged when stormy winds drive the turbid waves of high tides against the cliff; it emerges at low tide in a broad and gentle seaward slope, terraced near its top with long lines of horizontal benches or "curbs," which mark the reach of recent high tides. Where cobbles and pebbles are absent, the low-tide beach is at some places dangerous from quicksands; where storms have removed the beach detritus, the wave-cut rock platform which underlies it is laid bare.

Along the middle of the reëntrant in the northwestern coast, known as the Bight of the Somme, between Boulogne-sur-Mer and Dieppe, the upland margin is fronted for a distance of 65 k. by an alluvial lowland, 12 k. in greatest breadth, except where it is interrupted by the Somme estuary. The outer border of the lowland bears a belt of dunes, which broadens to the north toward Boulogne; the inner border is a line of abandoned sea cliffs, which must have been cut back by the sea before the low plain was formed, and which were then presumably as steep as the exposed cliffs at Dieppe and farther southwest, but which are now of more moderate slope, since the waves no longer remove the talus from their base.

The clift coast of the northwestern uplands does not favor navigation, as it has no good natural harbors. Many of the smaller upland valleys, without streams except in wet weather, are not cut down to sea level; they appear merely as depressions in the cliff top when seen from off shore. It is only in the larger valley mouths, drained by perennial streams and therefore cut down deep enough to be entered by short sea arms, that harborage is found; and even here the beach detritus is so actively and abundantly swept along shore by storm waves and tidal currents that it is difficult to prevent the harbors from being filled up. Long jetties have been constructed for this purpose where the need of harbors warrants it. Protection is thus afforded vessels at a dozen points; the chief ports where, with the aid of breakwaters and dredging, depth of water sufficient for larger vessels is provided are Calais (72,322) and Boulogne-sur-Mer (53,128) on the Straits of Dover (*Pas de Calais*), Dieppe near the mid-length of the clift coast, and Havre which lies on the north side of the largest reëntrant, a little back from the clift outer coast at the estuarine mouth of the largest river, the Seine. Etretat, Fécamp, and St. Valéry-en-Caux are watering places in valleys on the coasts between Havre and Dieppe.

CHAPTER XII

THE LOWLANDS OF NORTHERNMOST FRANCE AND WESTERN BELGIUM

60. *The Lowland, the Maritime Plain, and the Dunes.* The chalk uplands of Picardy and Artois fall off abruptly northeastward to the lowland of the district known as Flanders, partly in France, partly in Belgium, where the chalk strata, warped down to lower levels, are overlapped by younger beds of sand and silt. This area may be divided into three belts parallel to the shore line; the middle belt is a low maritime plain, about 30 k. wide in Belgium, but widening northeastward; it is of fine, moist soil, hardly above the level of the sea, from which it is separated by the outer belt of beach and dunes; the third belt is farther inland, a gently undulating sandy lowland, which rises gradually southeastward; it may be called the lowland of Flanders. The lowland will be first described; then the maritime plain, and finally the belt of dunes.

The Lowland of Flanders. The lowland is relatively infertile: it is the narrow beginning of a long belt of sandy country originally covered with heathery moors, which widens northeastward and stretches across Belgium and Holland into Germany. Parts of the surface are more fertile, where it is cloaked with a fine yellow loam, which is rapidly transformed into mud in wet weather, as many a soldier knows who has served in the trenches hereabouts. Occasional hills surmount the lowland. On the French side of the boundary, the picturesque town of Cassel, east of St. Omer and 28 k. south of Dunkirk on the coast, crowns one of the hills, 170 m. in height; to the east near Ypres, on Belgian territory, Mt. Kemmel rises to 150 m. altitude.

Farther inland, two broad areas of lowland plains, almost free from hills, lie at distances of from 50 to 70 and from 80 to 100 k. southeast of the coast at Dunkirk; the first plain is drained northward across the Belgian lowland by the Lys; the second is similarly drained by the Scarpe and the Escaut, small streams of leisurely flow, hardly below the level of the adjoining meadows. East of the second plain, the lowland rises toward the interior, with occasional hills above the general surface: thus 5 k. north of Tournai on the Escaut, Mt. St. Aubert reaches an altitude of 150 m., while the rolling lowland stands at 30 or 40 m. Many of the main roads hereabouts are surfaced with square-cut " Belgian blocks " of resistant rock, mostly from the Ardennes, thus making a durable but rough roadway.

It is into this lowland district that the belts of strongly tilted coal-bearing formations, which reach the surface farther east along the northern slope of the Ardennes (see map, p. 153), are extended through western Belgium into France (see map, p. 134) beneath the horizontal strata — chalk, clays, and sands — of the Paris basin, and as a result this district is of great industrial importance. All the many coal mines upon which the industries depend must sink their shafts as at Lens near the southern side of the Lys plain, through some 50 or 100 m. of barren horizontal strata, before the tilted coal beds are reached.

Lille (217,807), an important industrial and university city on the low plain of the Lys, is the largest of the region: Roubaix (122,723) and Tourcoing (82,644), manufacturing cities of rapid growth like Lille in the last century, are a little farther north in an indentation of the frontier, and across the frontier lie the Belgian cities of Courtrai a short distance farther on and Tournay to the east. Valenciennes is on the low plain of the Escaut; Cambrai is on the same river and Douai on its branch, the Scarpe; the last two lie near the southern border of the lowland near the chalk uplands on the south. All of these industrial cities profit from the coal mines of their district.

61. *The Maritime Plain of Flanders.* The maritime plain is of special interest. The fine soils of the plain are underlaid at a moderate depth by the seaward extension of the sandy strata which form the inland plain, and which testify by their marine fossils to an ancient submergence of the region beneath the sea. But between the surface soils and the underlying sandy strata, an extensive marsh deposit of peat and other fresh

FIG. 43. THE MARITIME PLAIN OF FLANDERS

water plants (black in Fig. 43) is found, a meter or more in thickness. A broad emergence of the region must have taken place in prehistoric times to permit the formation of the fresh-water marsh deposit. Then a submergence must have ensued, as the fine soils, which now cover the marsh deposit to a depth of several meters, contain shells like those still found in the adjoining sea.

Singular to relate, the upper layers of the marsh deposit contain relics of the stone age and the bronze age of prehistoric man, as well as Gallic and Roman coins down to the fourth century of the Christian era. Hence the submergence of the marshy belt presumably occurred in the next following century. After the submergence took place, the sea border of Flanders must have resembled the present Frisian coast of northern Holland, where the outlying dune islands are separated from the mainland by a debatable belt, alternately flooded and laid bare by the rising and falling tide; and just as those Frisian islands are now inhabited, although the debatable belt behind them is ordinarily submerged, so the dunes along the Flemish coast appear to have been inhabited through the dark ages from the seventh to the twelfth century, while the area of the maritime plain that they enclose was submerged under a very shallow sea. The later emergence whereby the present maritime plain was formed is believed to be due not so much to an uplift of the land as to the accumulation of marine sediments, swept into the shallow sea by currents from the southwest.

The maritime plain of Flanders thus constituted is remarkably level, hardly varying two meters in height in as many kilometers of distance. Its surface is somewhat lower than present sea level at high tide. It therefore offers a remarkable contrast to the chalk uplands of Picardy and Normandy. There water is deficient and the village houses cluster around deep wells; here water is in excess; the plain is not only threatened by floods from high tides, especially when landward storms brush the rising sea up on the shelving shore, but it is also in danger of floods from rain and from rivers.

The low and level plain must therefore be drained by many ditches and canals to the sluggish streams, the fields must be enclosed by dikes, and the stream mouths at breaches in the dunes must be closed with locks; the locks are kept shut at high tide when the sea almost reaches the base of the dunes, but are opened twice a day to discharge the streams when the tide falls and lays the beach bare far outside of the high-tide shore line. So threatening are the dangers of inundation from land, sky, and sea, that the inhabitants of the plain have for centuries past organized themselves in societies, known as "Wateringues," for maintaining the dikes, the canals, and the locks.

62. *The Dune-bordered Coast.* The dune-bordered shore line of Flanders is almost rectilinear and the sea is so shallow that

the retreating tide lays bare a broad flat of sands. All the ports of the smooth shore line are of artificial construction; their basins must be dredged and their entering channels must be protected by long jetties. Calais lies at the junction of the plain and the chalk uplands on the south, and is the chief port in French Flanders; its preëminence results from its nearness to England; but this element of its situation has exposed it to the hardships of repeated sieges. Then follows Gravelines, just back of the dunes at the mouth of the Aa, with St. Omer on the same small stream at the inner border of the maritime plain, 30 k. back from the shore line; next is Dunkirk, the chief port for the industrial cities of northernmost France, but formerly repeatedly fought over, as in the Battle of Dunes, when the French and English defeated the Spanish in 1658; only two hundred years ago it was the resort of so many corsairs who raided shipping that on the demand of the English its fortifications were demolished and its harbor filled up. The maritime plain here is narrow: the belt of dunes, with the city at the breach, is distant only 12 k. from the town of Bergues on a low hill at the inner margin of the plain.

The present residence of the King of Belgium is just across the frontier in the village of La Panne, which was saved from conquest in the autumn of 1914 by opening the locks at Nieuport at the mouth of the Yser, 12 k. beyond, and flooding a tract of the plain between the dunes and Ypres, near the inner border of the maritime plain. Farther on is Ostend, important in time of peace for its traffic across the Channel; then, opposite Bruges, 15 k. from the sea on the border of the inland plain, come the shore towns of Blankenberghe and Zeebrugge, the latter used by the Germans as a submarine base. A few kilometers farther on beyond the border of Belgium, the irregular estuarine coast of Holland gives access to the interior; there, 65 k. inland from the outer coast lies the great Belgian port of Antwerp (*Anvers*) on the Schelde; this city, although held by the Germans, does not give them lawful outlet to the sea, because the navigable waters below the city are in Dutch territory. Farther on beyond the mouths of the Rhine, Rotterdam is a leading Dutch port, 25 k. inland on a dredged waterway. Then the continuous coast begins again, with its long line of dunes back of the beach, and back of the dunes a broad repetition of the maritime plain of Flanders in the "polders" of Holland as far as the shallow Zuyder sea. Near the beginning of the stretch, just behind the dunes, lies The Hague (French, *La Haye;* Dutch, *'s-Gravenhage*), now famous as the seat of the international Peace Court.

The People of Flanders. Although the plain of Flanders is today shared along its length by France, Belgium, and Holland, its people vary more across its breadth, according as they live on the inland plain, the maritime plain, or the dunes, than according to their nationality. The infertility of the sandy inland plain has been overcome by the persistent industry of its hard-working population; farther in the interior, the surface is higher and undulating with a better soil; there the population is denser and more prosperous. The maritime plain has on the other hand a rich soil, the cultivation of which well repays the labor that it demands: the plain is therefore closely dotted over with the buildings of well-to-do farmers, whose dwellings are erected on artificial mounds a meter or more in height, with stables and sheds around them; the laborers live in villages, above which the church towers form the only landmarks. The people of the dunes are fishermen, sea-farers, or tradesmen, gathered in villages and cities at stream mouths; here the population is greatly increased in summer time by seashore visitors.

63. *The War Front in Flanders.* Military operations on the maritime plain of Flanders and on the inland plains of northernmost France have been hampered by the presence of ground water at a small depth beneath the surface, so that trenches are soon transformed into muddy ditches. Besides this, the whole surface of the plains becomes waterlogged in wet weather, making movement across the fields almost im-

possible. Thus the advance made by British and French troops across the maritime plain to the low hills on the border of the inland plain in August, 1917, was halted more by a heavy rain than by the resistance of the opposing forces. The difficulty of movement here operates against both armies.

The Germans have not succeeded in reaching Dunkirk and Calais, manifest objectives of great value as bases to prevent the transportation of British troops and supplies across the narrow Channel; the Allies have not yet reached the equally manifest objectives of Ostend and Zeebrugge, the possession of which would prevent the use of the harbors of the Belgian coast as bases for German submarines; for be it remembered that, as noted above, the Belgian port of Antwerp has access to the sea only through a waterway that lies in Dutch territory.

Of the two pairs of objectives, the second would appear to be the more difficult to maintain after gaining it; for until the Germans are driven altogether out of western Belgium, the attainment of Ostend and Zeebrugge would give the Allies only a low and narrow coastal belt, the occupation of which would be as difficult as that of the proverbially unfavorable position "between the devil and the deep sea." The extreme flatness of the maritime plain and the inland plains of the Lys and the Escaut tends to emphasize the strategic value of the low hills by which they are bordered: hence the necessity of dislodging the Germans from the semi-circle of hills east of Ypres, which the British forces gained in the summer and autumn of 1917; hence again where the chalk uplands on the south overlook the plain of the Lys, a desperate battle was waged for the possession of the outlying Vimy ridge, north of Arras and overlooking, though only from its moderate height of 124 m., the lowland occupied by Lens, an important coal-mining center. The occupation of this rich mining and manufacturing region by the Germans since an early stage of the war has been a heavy loss to France.

CHAPTER XIII

REGIONS NORTH AND NORTHEAST OF FRANCE

64. *The Ardennes and Beyond.* The uplands and highlands of southeastern Belgium, with small adjoining areas of France and Luxembourg and a larger area of western Germany, are known in their higher part as the Ardennes (see map, p. 153). Their total area roughly resembles a half-moon, measuring 180 k. east-northeastward along the diametral side, which is followed for much of its length by the Sambre-Meuse valley, and 80 k. across (southward) to the convex margin; the greater altitudes are from 400 to 580 m. If this region is approached from the south, between the Meuse and the Luxembourg frontier, ascent is soon made from the overlapping strata of the sixth upland belt of northeastern France, as described in section 35, to the highland of the Ardennes proper; on the west, where the ascent is more gradual, the lower slope is overlapped by the northern extension of the chalk uplands of northern France, as described in section 46.

The gradual northwestward descent of the Ardennes to the Sambre-Meuse valley is continued beyond it by the uplands of central Belgium; and these slope down to the lowlands of the coast and of the estuarine district of Holland. The northeastern part of the highland area declines through uplands to the lowlands of the Rhine west of Cologne. To the east, the Ardennes are adjoined by the Eifel highlands of western Germany; farther south the Ardennes are separated from the Hunsrück section of the Slate-mountain highlands (*Schie-*

fergebirge) of western Germany by the Luxembourg embayment of less altitude, into which the strata of the sixth and seventh upland belts of eastern France enter northeastward in tabular masses, as will be described in section 71.

The most striking feature of the Ardennes is the general evenness of their high-standing areas, which are sometimes so flat as to be boggy in spite of their altitude. The next most striking feature is the irregular course of the deep and narrow valleys by which the highlands are trenched. No account need here be taken of rock structure, partly because it is very complicated, more because the different rock masses, steeply inclined as a rule, are truncated without regard to their composition or attitude by the highland surface, and because the valleys very generally follow courses that pay no heed to the rock-formation boundaries. It is only by way of exception, and chiefly in the area near the mid-northern boundary of the region, that the upland ridges and the valleys between them show a relation to the northeast-southwest trends of the stronger and weaker structural belts.

Some of the valleys, especially in the less elevated uplands, are open enough for easy occupation; others, especially in the highlands, are narrow and steep-sided. Many of the rivers are extraordinarily serpentine; such is most conspicuously the case with the Semois, which flows westward in the southern slope of the highland (see diagram, p. 81), and to a smaller degree with the Lesse, which flows westward from the central highland area; both of these rivers flow into the north-flowing Meuse in its narrow, gorge-like valley, which is also sinuous, though not to so remarkable a degree as that of the Semois. Further east, the Ourthe and its main branches, the Amblève and Vesdre, surprisingly sinuous in much of their course, drain a large highland area northward to the second elbow of the Meuse at Liège, where it turns north toward Holland.

In the higher and sharply incised districts, the valley sides are wholly abandoned to forest growth, although parts of the highlands themselves are cleared and cultivated: thus the town of Rocroi (390 m.) stands in the midst of a farming district on the upland west of the Meuse gorge; its church tower may be seen from the high-

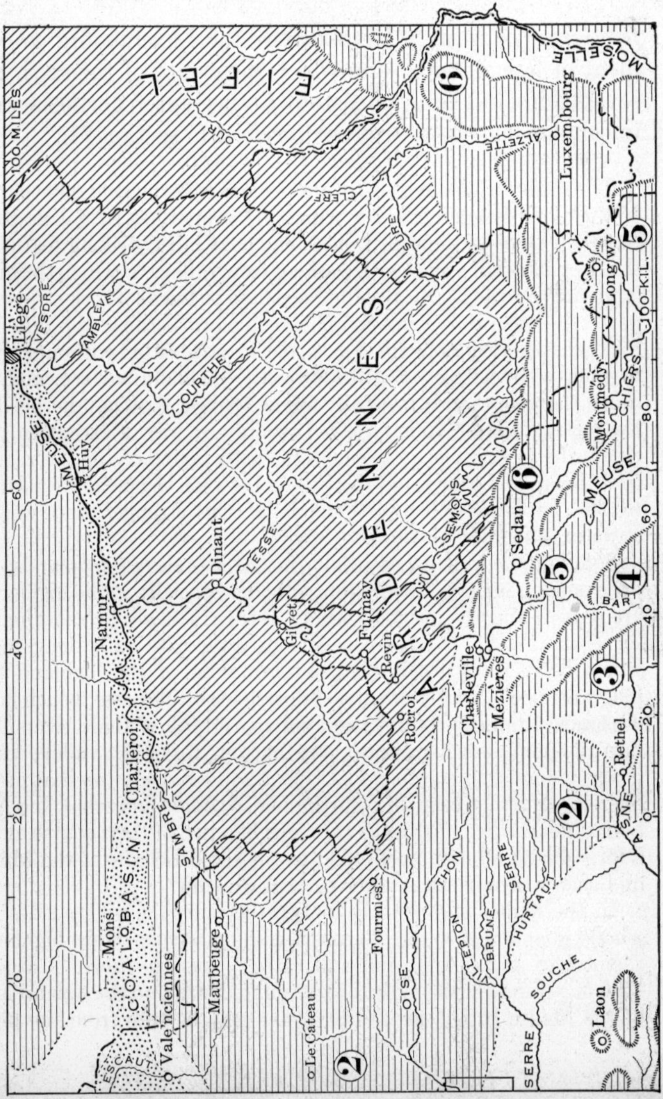

FIG. 44. THE REGION OF THE ARDENNES

lands miles away to the east, rising over the even skyline, with the deep gorge of the Meuse hardly perceptible in the middle distance. Although the region is by no means mountainous, and although movement is easy upon the highlands between the valleys, the Ardennes are difficult to traverse because of the deep, narrow, and sinuous valleys by which they are trenched. Together with the Slate-mountain highlands to the east, they separate the path of the German invasion of 1870 from the path of the invasion of 1914 by a distance of 250 k.

65. *The Gorge of the Meuse.* The narrow and winding gorge of the Meuse is the only cross-cut by which the traverse of the Ardennes highlands can be avoided. It would be called a canyon in the western United States. Its situation is peculiar, for the river in adopting its course through the high ground instead of running westward around it seems to have defied the general rule that water runs down hill. In explanation of this peculiarity it may be confidently believed that the course of the river was adopted antecedent to the upheaval of the highland, when the Ardennes area was lower than the basin of the upper Meuse in northeastern France; further, that the upheaval of the highland was so slow that it did not suffice to turn the river out of its antecedent course: the river cut down its gorge about as fast as the highland was upheaved. Rivers of this kind have been aptly compared to a bandsaw in a lumber mill; the saw holds its place while ripping its way through a log that is pushed against it.

The gorge of the Meuse is occupied, in spite of its narrowness, by several towns which take advantage of local widenings of its floor: thus Revin, Fumay, and Givet lie within French territory, and Dinant in Belgian. The adjoining cities of Mézières and Charleville have already been mentioned as lying in the open country to the south on the loops of the river before it enters the gorge; and Namur as marking the exit elbow, where the Meuse is joined by the Sambre from the west: this river also follows an incised meandering valley, but its depth is much less than that of the Meuse gorge.

It was in the salient angle limited by the gorge of the Meuse on the east and the valley of the Sambre on the north, with the fortified city of Namur at its apex, that the French Fourth Army and the British Expeditionary Force attempted to check the German advance in August, 1914; but Namur was soon captured and the Germans, sweeping westward over the more open country north of the Sambre, threatened the British so seriously that the Allied forces were compelled to begin a retreat that was not stopped until the battle of the Marne was fought, a fortnight later, 180 k. to the south.

An important railway runs with the Meuse through the Ardennes; but as it is dominated by the adjoining uplands, which for half the length of the river gorge lie in a northward loop of the French frontier, and as the fortress of Mézières lies near the gorge entrance, this railway was not used by the Germans for the transportation of troops until after their army had passed around the western end of the highland area, and the retreat of the Allies had left the Ardennes open to enemy occupation.

66. *The Uplands of Central Belgium* decline slowly from the Sambre-Meuse valley toward the coast and the lowlands of Holland; they are underlaid by the chalk and younger strata which have wrapped around into Belgium from the broad northwestern saddle of the Paris basin. The upland surface is worn down in wide-spaced, irregularly branching valleys, whereby it becomes broadly undulating: it is everywhere open to easy movement and the greater part of it is rich agricultural land, supporting a large population. A less productive sandy strip slopes gradually to the maritime plain of Flanders, as described in section 60.

The valley of the Sambre-Meuse along the northern base of the upland and highland area here considered is unlike the winding valleys within the highlands, in that it follows in a general way an almost rectilinear east-northeast course, due to the structural guidance of certain steeply inclined weak strata in the long and narrow belt of the Belgian coal basin. The coal basin, if followed southwest, leaves the upper Sambre valley and turns westward under younger covering strata into northern France, as stated in section 60; in the opposite direction beyond the northward turn of the Meuse valley at Liège, it extends northeastward into western Germany; farther on in the same direction and across the Rhine lie the great industrial districts centering at Essen, the seat of the Krupp works in Rhineland, and at Dortmund in Westphalia.

In the districts where the coal beds are most actively exploited, as in the western part of the Belgian field, the mining villages are connected by a network of railways. The coal-bearing belt is occupied or adjoined by many industrial cities; Valenciennes, Douai, and Lens in France have already been mentioned: Mons, Charleroi, Namur, le Huy, and Liège (German, *Lüttich*) lie in Belgium, and Aix-la-Chapelle (German, *Aachen*) is in Germany. Hence on rounding the western slope of the Ardennes, the Germans not only made entrance into the open country of northern France, but took possession of a highly productive industrial region on the way, which they have since worked to their own advantage.

The uplands and lowlands of northern Belgium are drained chiefly by the Escaut (Schelde) and its branches, of which the chief are the Senne and the Dyle (see map, p. 10); on these two streams lie Brussels (French, *Bruxelles;* German, *Brüssel*) and Louvain (German, *Löwen*), about half way from Charleroi and Namur to Antwerp (French, *Anvers;* German, *Antwerpen*), the chief Belgium port, where the Schelde widens as an estuary. This region has been repeatedly fought over in earlier centuries: Waterloo, the most famous battlefield of the first half of the nineteenth century, lies on the upland 15 k. south of Brussels.

It should be remembered that a southward arm of Holland, including the city of Maastricht (H, Fig. 13), extends along the Meuse (Maas) and brings the Dutch frontier to within 15 k. of the northern border of the Ardennes, where the highland slope is deeply dissected by the Ourthe and its sinuous branches: hence the German invasion of Belgium was confined to this narrow space. There at the elbow of the Meuse lies Liège, which consequently had to bear the brunt of the first attack.

67. *The Lorraine Plateau and the Adjacent Districts of Germany.* The upland saddle that spreads eastward from the Paris basin between the highlands of the Vosges and the Ardennes passes from the part of the old province of Lorraine which is still retained by France to the part which, under the name of Lothringen, has been German territory since 1871, and beyond (see map, p. 54). It is an upland area, rather sharply incised by many irregularly branching valleys. Along its northern border is a belt of hillls, 30 or 40 k. in breadth and 100 k. in length, and beyond these rises the Hunsrück section of the Slate Mountains. The eastern border of the plateau falls off to the lowland plain of the middle Rhine by a well-defined escarpment, due to a displacement on a profound fracture of the earth's crust. The name, Vosges, is sometimes extended to a southeastern part of the upland, from 350 to 450 m. in height, overlooking the Rhine plain, next north of the highlands of the Vosges proper: farther north, a more elevated forested area (570 m.) along the border of the upland, which there advances further eastward, is known as the Hardt (or Haardt), in the political province of Pfalz.

The less elevated upland, lying west of the Hardt and including the hilly belt south of Hunsrück highland, will here be described with the Lorraine plateau. The chief rivers of the plateau are the Moselle (German, *Mosel*) which crosses

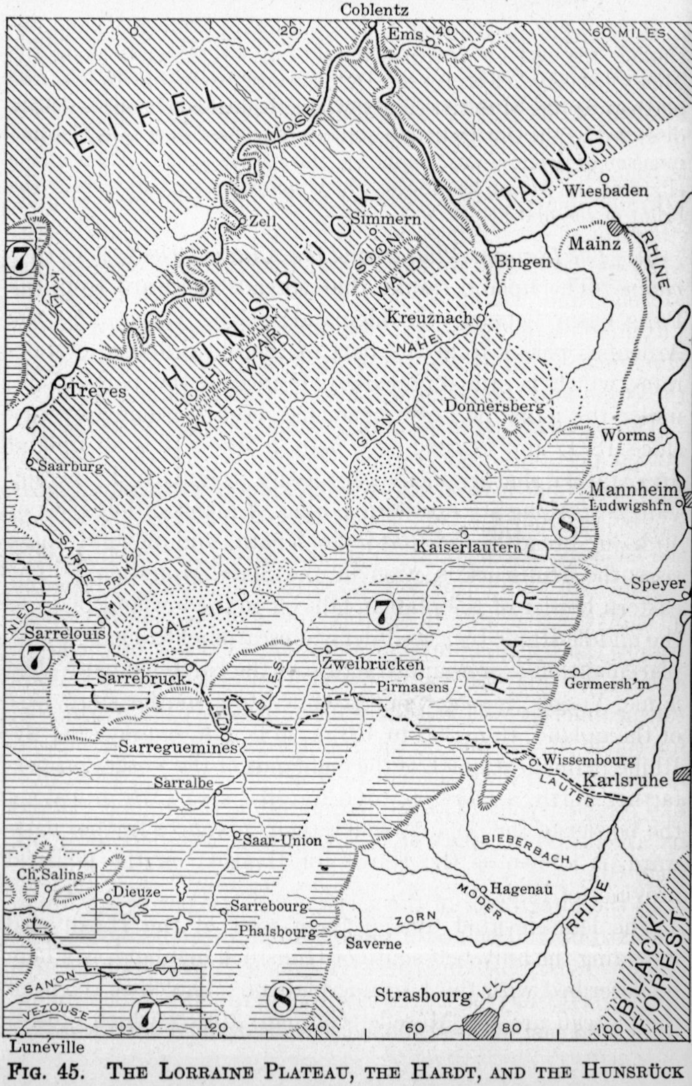

Fig. 45. The Lorraine Plateau, the Hardt, and the Hunsrück

its northwestern part, and the Sarre (German, *Saar*), which rises in the Vosges and flows north-northwest across the plateau to the Moselle. These rivers and their branches flow in valleys that are well incised below the upland level; more is told of them below. An upland area, associated with the Lorraine plateau and covered by a long lobe of the limestones of the seventh and sixth upland belts, extends northward between the Hunsrück and the Ardennes west of the Moselle, and will be described below under the name of Luxembourg embayment; the hilly belt next south of the Hunsrück, occupied by tilted ancient rocks and drained by the Nahe and the Glan, will also be treated on a later page.

The eastward view of the Lorraine plateau from the detached portions — the Grand Couronné — of the fifth upland belt (400 m.) on the French side of the frontier near Nancy, or from the bold scarp of the same belt (350 m.) on the German side of the frontier near Metz (see map, p. 60), discloses an undulating landscape (300 to 350 m.) stretching 60 or 80 k. eastward. In the foreground of the view from near Nancy the irregular tabular extension (350 m.) of the sixth upland belt may be distinguished where the frontier follows the meandering valley of the Seille, a tributary of the Moselle, for some 30 k.; one of the tabular masses stretches 20 k. northeastward into German territory in a long promontory-like forested spur (330 m.) surmounting the rolling upland by about 100 m.; part of it is shown on the detailed map, page 161.

In the foreground of the eastward view from above Metz, the sixth upland belt is of more normal form; its moderately scalloped scarp (310 m.), here lying some 20 k. east of the fifth upland belt, may be traced northward for 80 k. The rolling lower land east of the scalloped scarp is drained northward by

Fig. 46. A Forested Spur of the Sixth Upland

Belt advancing Northeast of the Frontier

the French Nied, so-called to distinguish it from a more eastern branch, the German Nied, of the same trunk stream, the Nied, which runs northward to the Sarre.

68. *Southern Part of the Lorraine Plateau.* If the southern part of the plateau is now crossed eastward, a view from any one of its many broadly rounded hills between the valleys of the Seille, the Sanon, the upper Sarre and their ramifying head-waters, shows a gently undulating skyline in which all the other hill crests unite. The skyline thus represents the upper surface of the broadly extended limestones that determine the seventh upland belt, the southern part of which has been described in section 25. The limestone area stretches eastward across the Sarre and ends in an irregular margin (350–400 m.), overlapping upon the underlying sandstones which rise southeastward in the mountainous hills of the Vosges; the same sandstones also extend eastward with small increase of altitude for 15 or 20 k. to the scarp where the upland that they maintain descends abruptly to the Rhine valley. The soil of the limestone uplands is fertile and forests are of restricted area upon them; the sandstones have an infertile soil and are largely forested.

The main lines of travel and transportation—road, canal, and railway — eastward from Paris and Nancy pass across the southern part of the Lorraine plateau, next to its ascent into the higher Vosges, on the way to Strasbourg, the chief city of Alsace in the Rhine valley-plain, and beyond. The national road is literally a highway east of Nancy, for though it must descend in crossing the middle course of the Seille near the frontier and headwaters of the Sarre farther east, it avoids the valleys as far as possible and follows the broadly arched hill crests of the limestone area; continuing, it reaches the very extremity of a sandstone spur, 428 m. in altitude, directly overlooking the lowland of the Rhine, to which it descends in zigzags at the mouth of the short valley drained by the Zorn. The canal lies more to the south and follows the valleys: it first ascends the Sanon, a branch of the Meurthe, beyond the frontier; then crosses the divide to the upper valleys of the Sarre and winds around the hills that enclose them; here a branch canal turns north down the Sarre; the main canal then tunnels 2 k. under the next divide, and descends through many locks 100 m. in 13 k. along the valley of the Zorn to the Rhine lowland. Some of the valley heads near the upper Sarre are occupied by canal reservoirs, 3 or 4 k. in length, which branch into smaller side valleys above their dams. These small water bodies and the canals that they serve are the chief barriers to movement hereabouts.

The main line of the Eastern railway, after making a southward detour from Nancy to Lunéville on the Meurthe, turns up the Vézouse through the open country west of the Vosges (see section 25), but soon leaves the stream, crosses the frontier midway between the Vézouse and the Sanon (see the detailed map, pp. 60–61), and then crosses the upper Sarre and tunnels the divide alongside of the canal to the Zorn valley. All three lines unite below the escarpment at Saverne (German, *Zabern*) on the Zorn, a town that became notorious from a typical militaristic incident, the affront of a civilian by an army officer, the year before the war; it was in this case that the decision of a civil court was reversed by the military authorities.

69. *Northern Part of the Lorraine Plateau.* A traverse of the northern part of the Lorraine plateau discovers a greater variety of relief than that of the southern part, for the valleys of the Sarre and its branches are here more deeply eroded; moreover, the foundation rocks, which occupy the hilly belt farther northeast and the highlands of the Hunsrück and the Eifel, are here first seen in an area on the Sarre where the covering limestones and sandstones of the Lorraine saddle have been worn away. It is interesting to note that along the course of this river seven towns embody its name in theirs, the first and last being essentially the same.

Around this area of the foundation rocks, the limestones of the seventh upland belt, overlying the basal sandstones, appear in a well defined upland rim with a scalloped scarp; and within the rim, among the deformed foundation rocks, lies a coal field (dotted on map, p. 158) named from the Sarre (*Saar*) which flows across its western end, and on which Sarrebruck (German, *Saarbrücken*) and Sarrelouis (German, *Saarlouis*) are important coal-mining centers. The district is of great economic importance, as it furnishes a large supply of coal, some of which is used in the iron furnaces on the Moselle below (north of) Metz, for smelting the minette ore from the fifth upland belt. This district is furthermore historically interesting, as it was at Sarrebruck on German territory that the first encounter of the Prussian and French armies occurred in July, 1870.

The limestone uplands continue to the east of the coal field through a district known as Westrich, where they are drained by the Blies; their greatest altitude (450 m.) is reached about 50 k. northeast of the Sarre, where their margin is extremely irregular with many outlying patches. Farther on, the underlying sandstones yield an infertile soil and are generally forested; here the upland gradually rises to the Hardt (500 to 600 m.); the scarped border of the upland lies some 30 k. beyond the outlying patches of the limestones, and as it trends northeastward, it here stands from 30 to 50 k. farther east than the upland border near Saverne. Both the limestone and the sandstone uplands are deeply cut by many irregular and close-set valleys, making the district difficult to traverse. The chief towns of the limestone area are Sarreguemines (German, *Saargemünd*) and Zweibrücken. Pirmasens and Kaiserslautern lie on the adjoining sandstone area.

The boundary between France and Germany before the war of 1870 crossed this region in an irregular eastward course from a sharp bend of the Moselle, marked by the town of Sierck about 10 k. below its emergence from the sixth upland belt (see map, p. 67), to the Rhine lowland. It passed south of the Sarre coal field along the Sarre and the Blies through the Westrich district, and across the southern part of the Hardt; it then descended to the broad valley lowland a little north of a southward salient or prong in the marginal scarp, known as Hochwald, and followed the Lauter to the Rhine: Wissembourg (German, *Weissenburg*) lies at the base of the scarp next south of the line. The boundary between Lorraine and Alsace ran southward along the scarp into the Vosges.

The Hilly Belt south of the Hunsrück. To the north and northeast of the Sarre coal field, the foundation rocks there exposed constitute an irregularly hilly district of unsystematic form, except that occasional ridges manifest a northeast-southwest trend in accordance with the trend of the tilted belts of resistant rocks that determine them. The general altitude of the hills is from 350 to 550 m. Near the eastern end of the area, a mass of extra-resistant igneous rocks rises in the Donnersberg to 687 m. A small part of the area is drained westward by the Prims to the Sarre; the larger part is drained by the Glan and Nahe to the Rhine at Bingen. The lowland of the Rhine, east of the upland scarp, is of low relief, and densely populated. The important cities of Karlsruhe, Speyer, Mannheim, Worms, and Mainz lie on or near the river.

70. *The Hunsrück.* A gradual northward increase of altitude leads from the Lorraine plateau to the Slate-mountain highlands of western Germany. The highlands are divided into four subequal parts by the gorge of the Rhine, trending northwest, and by the gorges of its opposing tributaries, the Moselle from the southwest and the Lahn from the northeast. The southern part is known as the Hunsrück, the western part as the Eifel.

The Hunsrück is a rolling highland, measuring 100 k. parallel to the Moselle, and 40 k. in a transverse direction; its general altitude is from 450 to 500 m.; but it is surmounted by several linear forest-clad ridges known as the Hochwald Idarwald, and Soonwald, trending northeast-southwest, and reaching heights of from 600 to 800 m.; and it is deeply cut by narrow steep-sided valleys around its borders toward the Sarre which flows to the Moselle on the southwest, the Moselle on the northwest, the Rhine on the northeast, and the Nahe which flows to the Rhine on the southeast. The undulating highland surface, away from the surmounting ridges and back of the marginal valleys, is cleared and cultivated, and is the seat of many quiet, out-of-the-way villages. All the bordering valleys are followed by roads and railways; the highland is traversed chiefly by roads.

71. *The Luxembourg Embayment.* The Eifel resembles the Hunsrück in being a rolling agricultural highland, sharply dissected by deep-cut and frequently meandering valleys around its margin; but it has few surmounting eminences It is confluent westward with Ardennes, except that on the south the two highlands are separated by the somewhat lower uplands of the Luxembourg embayment. This embayment is structurally similar to that by which the limestone strata of the seventh upland belt are led from the Lorraine plateau to the Hardt; but here two upland belts, the sixth nested in the seventh (see map, p. 153), are led northeastward, and the Luxembourg embayment is therefore of more complicated surface form than the Lorraine plateau; but on the other hand, the troughing of the strata is here more pronounced than in Lorraine, and the scarped edges of the controlling strata, those of the sixth belt better defined than those of the seventh, have more regular trends. Their description would therefore be

comparatively simple, were it not that they are obliquely cut across by the deep and winding valleys of certain northern tributaries of the Moselle; as a result the verbal description of the separate members into which the upland belts are divided can hardly be attempted here.

A brief statement must suffice. Let it first be understood that the sixth upland belt trends about parallel with the northward course of the fifth upland belt (see section 32) where the Moselle flows between the two from Metz to Thionville; but then, instead of soon turning to the west like the fifth upland belt north of Thionville, the sixth extends 60 k. to the northeast into the Luxembourg embayment, from which it returns sharply to the southwest before it again parallels the fifth belt, trending westward, along the southern border of the Ardennes, as shown on page 81. The slender northeastern point of the sixth belt (350 m.) is divided into several distinct hills; farther southwest, its two scarps, one facing southeast (350 m.), the other northwest (400 m.), are cut across by the Sure (German, *Sauer*) and several branch streams which rise in the Ardennes and flow through deep, narrow, and sinuous valleys southeastward to the Moselle. Farther southwest still, a long reëntrant is cut back in the northwestern scarp, at the head of which lies the city of Luxembourg, capital of the Grand Duchy of the same name. Thence, westward, the sixth belt is fairly well developed on the flanks of the Ardennes, 20 or 10 k. north of the fifth; but its continuity is interrupted by the notches of many streams flowing southward from the highlands to the Chiers or the Meuse, and its border is not marked by a distinct scarp, as section 35 has already made clear.

The seventh upland belt, less distinct than the sixth, makes a larger and broader excursion to the northeast: the apex of its curve lies 80 k. northeast of the turn of the fifth belt by Thionville, and 20 k. beyond the extremity of the several hills of the sixth; its imperfectly developed scarps are turned southeast toward the ascent of the Eifel highlands and northwest toward the ascent of the Ardennes highlands: like the sixth belt the seventh belt and the adjoining highlands also are frequently cut across by deep valleys; not only by those of the Sure and its branches, but also farther to the northeast by the Kyll. On returning from the embayment along the slope of the Ardennes the strata of the seventh upland belt are overlapped by those of the sixth and disappear: this foreshadows the fate that, further to the west, overtakes the strata of the sixth, fifth, fourth, and third belts in turn, until the last of them disappears under the northward overlap of the chalk near the headwaters of the Oise on the southern slope of the Ardennes west of the Meuse, as shown on page 153.

72. *The Gorge of the Moselle.* The course of the Moselle northeastward from Thionville, where it was left in section 33, soon leads it obliquely through the sixth upland belt, thence northward along a valley that is incised in the lower land east of the upland scarp, then again northeastward, to a belt of steeply inclined, weak strata, included between the more resistant rocks of the Hunsrück and Eifel highlands on either side, and extending for 55 k. northeastward or half of the distance to the Rhine; an open valley, from three to six k. in width, has there been excavated. The serpentine Sarre joins the Moselle from the south as the open valley is reached; the ancient city of Trèves (German, *Trier*) lies in the valley a little farther northeast. Singularly enough, although the open valley continues northeastward as far as its belt of weak rocks extends, the Moselle does not follow it so far, but turns to the right and cuts a gracefully serpentine valley in the border of the Hunsrück highlands for the rest of its way to the gorge of the Rhine. Through this part of its course it is joined by the many small streams that have cut their steep-sided valleys in the margin of the adjoining highlands.

Travel through this region is easy, though somewhat circuitous, along the main valleys; no serious obstacles are met on the rolling highlands, if the surmounting ridges and the incised marginal valleys are avoided; even the ridges may be ascended or crossed without especial effort, for their forested slopes, not over-steep, bear many paths, and their crests are occasionally topt with outlook towers from which repaying views are obtained. But the spurs of the highlands, advancing into the labyrinth of sharply incised marginal valleys, are traversed with difficulty, and should be avoided except for the pleasure of scrambling in a rough country.

The Gorge of the Rhine, a famous example of a narrow, river-cut valley across an uplifted land mass, offers a striking contrast to the quiet highlands on either hand. Busy river-side villages occupy strips of alluvium at ravine mouths on one bank or the other; terraced vineyards clothe the sunlit slopes; railways and, for most of the gorge-length, roads also follow both banks. The river channel has been cleared of the rocks that here and there made its navigation difficult, and steamboats for passengers and towboats with strings of barges for freight pass rapidly downstream, or more slowly upstream. A view of the gorge from an upland spur shows it to be an artery of human life, throbbing with activity, while the uplands on either side are the abode of a quiet rural population. May this famous view or the broader view from the margin of the Hardt across the populous Rhine valley further south be enjoyed, without undue delay, by many readers of these pages!

INDEX OF PLACE NAMES

PRINTED AT THE HARVARD UNIVERSITY PRESS, CAMBRIDGE, MASS., U. S. A.